Praise for *Build Bette*

"Over the years, I've met so many individuals in so many organisations who have heard about Agile and DevOps ways of working, and are inspired and enthused, yet nervously cautious about getting started—because they've never done it before. This is why Kristian's book is so important. It's like he's sat beside the reader, walking step-by-step through what it takes to launch an automation initiative and see it through to its successful conclusion, having done it a ton of times before. His voice has clarity and explains complex concepts in the language of a calm and friendly colleague who's lived it. He shares his own personal experiences, good and bad, with insights into his own career; his optimism and faith in the human spirit are palpable. The book is packed with tools Kristian has generously included which provide pragmatic and concrete examples of how to get this work done—and they don't just include ways to implement technology; there are plenty tips around how to influence stakeholders and behaviour too."

—Helen Beal, Chief Ambassadeur, DevOps Institute

"I highly recommend this book for anyone who wants a humanistic angle on the processes required for implementing automatic software deployment."

—Allan Kirkeby Andreasen, CEO, A2Sourcing.com

"*Build Better Software* offers sound advice on why and how to treat light-weight processes, automation, software, and code quality as first-class citizens and use them as pillars to build compelling software products."

—Torben Falck, Software Craftsman, Angel Investor,
Founder of Kwanify.com

"Describes an activity in the development of software that is too often neglected or rendered wrong.... It will reward the reader in the importance and best practices automating the build and deployment of one's digital solutions and increasing the quality of one's digital delivery."

—Pedram Shashavand, Partner, Hybrid Professionals

"This book is attacking the technical challenges from my favorite angle: It is all about people! While laying out a practical approach to the technical journey, the message is clear: Good technical solutions take into account the effect on people—and if you want to create real change start with building support and relationships."

—Therese Hansen, Scrum Master, Grundfos

"If you plan on putting DevOps into practice on a team, Kristian presents hard-learned lessons from doing the actual work—not just a theoretical assumption of the real world.

—Kenneth Christensen, Lead Software Architect, LEGO Group

"Kristian does an excellent job of taking the reader on for a journey in the world of CI/CD. With many great examples of what to do and what not to do in a lot of well described and well known contexts, the reader is given a good recipe for getting code from a developer workstation to production servers in the pragmatic spirit of DevOps."

—Casper Englund, CTO, MainChef

BUILD
BETTER
SOFTWARE

BUILD BETTER SOFTWARE

How to Improve Digital Product Quality and Organizational Performance

Kristian Bank Erbou

Foreword by Jayne Groll

BuildingBetterSoftware ApS

Smørhullet 6
7100 Vejle, Denmark

First Edition

Cover design and book design by Kelley Dodd

ISBN: 978-1-7356806-0-6
eBook ISBN: 978-1-7356806-1-3
PDF ISBN: 978-1-7356806-2-0

For information about special discounts for bulk purchases or for information on booking authors for an event, please visit our website at www.buildingbettersoftware.com.

Contents

Acknowledgments

THIS BOOK IS MY DEBUT in writing anything at length, apart from school work and studies, where the "audience" so to speak would never be more than one or more examiners collaborating to provide me with an outcome in the form of a grade. Nothing else would likely happen after that. The writings I had made would have fulfilled their purpose and been left forgotten in a cloud drive somewhere. It has proven to be an entirely different exercise to write a book that will be judged and put under scrutiny by potentially millions of people around the Earth—some of whom may be much smarter and wiser than me on exactly the topics that I perceive myself to also be somewhat of an expert in.

I have come to the conclusion, though, that social status is nothing more than perceived status. It has nothing to do with what you actually do know or are able to achieve; it's all a mindset thing. I believe, too, that lots of people shy away from creative artwork such as painting, writing, or cooking a homemade meal, because they put themselves on the stage in unpleasant ways. I can understand the mechanisms of not daring to try because you don't feel worthy of participating in a conversation with those who are perceived as being much smarter, much wiser men and women. I have felt the same for quite some time, but I have now decided in favor of putting myself at risk anyway. I may not live up to the expectations of some; that's a price I'm willing to pay if I can help others achieve insights and become a success when they might not have done so otherwise.

I couldn't have come to this conclusion without people and family around me who have engaged with me personally as well as professionally. Thank you. Without the support, the discussions, the solutions, the ups and downs in our pursuit of finding meaning to it all, this book would for sure have looked entirely different. A special thank you to all respondents to surveys and to reviewers —your feedback was invaluable in the process. To my wife and three kids: Thanks for not taking me too seriously along the way.

Foreword

DIGITAL DISRUPTION LEADING TO DIGITAL TRANSFORMATION is a mandate for just about every business these days regardless of vertical market. There is little doubt that every organization should be actively looking to increase its internal use of automation in order to increase its external ability to serve its customer. Born from the rise of adoption of frameworks such as Agile, DevOps, and Site Reliability Engineering (SRE), as well as the acceleration of artificial intelligence and robotic process automation, the ability to "automate everything" is within reach regardless of physical environment (cloud or on premise).

The pervasive use of automation is particularly important as more and more organizations accelerate their time to value, and build and release software in small, frequent increments. Enterprise interest in continuous delivery and deployment has grown rapidly over the past ten years as DevOps adoption has gained global attention. SRE is capturing the attention of the enterprise with a defined role and a directive to remove toil through engineering. Yet despite the focus on automation as the key enabler of DevOps, the organizations with the most success are those that have invested in the culture as much as investing in the tools.

This book provides actionable, concrete technical advice for building and deploying software faster by optimizing automation capabilities. While advocating for the automation of all infrastructure build-and-deploy processes, Mr. Erbou clearly recognizes the importance of humans in the automation equations. The human elements of critical thinking, process engineering, skilling, and new ways of working are deeply embedded into almost every aspect of the guidance. The result is a concrete model where humans and computers form a symbiotic relationship—each with its own unique native abilities, and each unable to proceed into the future without the other. The approach taken here very much aligns with the principles of Agile, DevOps, and SRE: make tomorrow better by automating your way out of manual work or toil.

If you take away anything from the upcoming chapters, remember this: you should automate everything involved in infrastructure build and deploy processes. You should also remember that automation is not your "why" or your organization's "why," according to Simon Sinek's definition of the Golden Circle. Automation is necessary to fulfill your "why" as we move into the new normal of the next decade.

— Jayne Groll, CEO, DevOps Institute

Preface

I AM A FUNDAMENTALLY NAÏVE SOUL who believes in the best in people. I believe, for instance, that people get up every day and go to work with the attitude that they will do their very best for the good of the company. It makes no sense for me that people would get out of bed and think, "Today I'm going to ruin somebody else's day at work!" No, I firmly believe instead that you and your colleagues across departments go to work with the attitude that you all want to be good corporate citizens.

In this context, the challenge from a leader's perspective is to ensure that all departments and individuals are pulling in roughly the same direction. Corporate culture and values must ensure that everyone has an intuitive understanding of what "the best" for the company is. However, despite best intentions, even the best values are challenged in daily life, where people attempt to reformulate conflicting interests and individual interpretations of abstract sets of values for the company into concrete behavior.

The classic issue I've seen many times as a software developer is the friction between a development department and an operations department, where both sides agree on mission statements like "We want to deliver quality to our customers." Reshaped into practice in departments and in individual teams, this mission statement becomes a matter of measuring how many new features developers can make in the shortest possible time, while the operating department will be measured by its ability to deliver stable operations on the infrastructure.

This kind of incentive structure in a company is poison for good cooperation because you build in friction when apartments have conflicting interests. It often happens, though, when missions aren't communicated properly and where management hasn't figured out that software development and software operations are learning processes based on empirical evidence and continuous mitigation of risks.

Recent frameworks called Agile and DevOps aim at bridging the gap between these conflicting interests and advocate for software development being a fact-based and holistic collaboration in the development of software and the systems on which the software runs. It should also be a unified approach toward developing a culture of collaboration and learning meaning. Both Agile and DevOps state that it is a mistake to organize your business so that different departments only deal with parts of the lifecycle. Without strong management mitigating organizational friction, there is an inherent risk of misalignment between development and operations departments in terms of culture and interpretations of what constitutes "good quality." If leadership does nothing, eventually both recruitment and incentive structures in various departments will end up supporting an organizational culture that focuses narrowly on their own interests without further consideration of the individual end-user experience.

DevOps is often equated with automation. That's partially right, but automation is just one tool in the toolbox for ensuring a higher quality end product. Like Agile, DevOps is more about the willingness to engage in relationships and collaboration across organizational units in a company—hearts and minds, if you like.

Automation gives the option of making a decision based on your desire to achieve something—lower costs, higher revenues, less risk, or similar. It may also be the tool that, when implemented properly, helps build a culture of collaboration across departments to achieve shared goals. It would be very easy to say that automation is the ultimate solution to all problems, but of course this is not the case. An investment in automation is a completely necessary tool in the toolbox if you want to work more efficiently, but it needs to be put into a context. Here the solution is often very different from company to company, from department to department, from team to team, and right down to the individual level. The solution for any problem varies greatly from organization to organization, even if the problem domain is basically the same, such as how to apply DevOps or Agile in an organizational context. Your company's long-term strategy—as well as your own personal skills, motivations, resources, and personality—are all variables that you must consider and weigh according to your desired end goal when searching for concrete methods and the answer to "What do I do now?"

This is not the book for those who want to learn the story of DevOps, about DevOps culture, about agile software development, about the latest trend in artificial intelligence, or how to use anomaly detection to automatically find and remove servers in one's infrastructure before they stop working as intended. You also will not learn how to set up build servers or whether to choose cloud-based solutions for your upcoming automation initiative. There is a sea of material on all these topics already.

I imagine you're reading this book because you're looking for answers to the question of how to automate promoting source code to production, or you'd like to get inspiration when opportunities open up by looking into an investment in automation of manual processes. This may be the book for you if you find yourself asking questions like, "What activities should we build into our automation of digital asset deployments? Give me a template!" You may be a software developer or have IT operations as your primary responsibility. You might also be a leader with current HR responsibilities and a bit of technical background, so you understand the nature of the everyday problems your team has.

Your background, however, is not that important to me, nor are your motivations, skills, or desires for the future. What I want to achieve with this book is to give a human being like you concrete guidance on how to motivate your managers and other stakeholders to invest in automation and the infrastructure and skills it brings, while also providing a blueprint of a viable plan of execution. My dream is that by reading this book, you end up with a clear picture of what you need to do to increase the quality of your deployments by introducing automation principles into your work. I want you to be using some of the basic principles of Agile and DevOps in your daily lives.

My hope is to convey all my experience with fifteen years of software development and Continuous Integration and Delivery in a form that enables you to build on my failures, successes, and learnings. You want to know how you get started. You are already well on your way, and this book will hopefully help you in your quest.

Part 1

The Great Why

IF YOU CAN—and it makes sense—you should automate everything related to build and deployment of your codebase in your infrastructure. Nothing prevents you from automating all processes in your entire infrastructure, even though it's located on-premise. The servers in your on-premise datacenter are no different than the servers in a datacenter at Google or Microsoft. The only difference is ownership and the fact that your servers are configured based on your needs and premises. Every physical element is in principle the same on-premise as it is in the cloud. If all stakeholders agree, you may very well program and automate your way out of creating servers, databases, database instances, firewall settings, network segments, and so on in your on-premise datacenter. It's perfectly possible to automate the installation of all necessary software artifacts, security patches, and configuration of software serving customer requests.

You may automate the installation of cryptographic certificates used to secure your systems from man-in-the-middle attacks, both self-signed as well as purchased ones. You may automate users creating on your databases against a password manager and automate access to your password manager. This would allow your developers to program access to systems and service in your infrastructure into their deliveries without ever knowing actual usernames and passwords because they are shielded from that knowledge by design. You may apply Configuration Management principles throughout the entire lifecycle of your software deliveries so employees will never need to log onto a server again. All state changes could be managed by an enterprise tool that has a complete history of all changes on all servers in every corner of your infrastructure going backward in time.

Everything is possible. The only physical laws that apply in software development are the physical laws related to the hardware on which the software is being executed. The speed of light dictates how fast electrons can make a zero into a one. The hardware itself also has an upper limit to, for example, how much disc space is available and how much data can move from A to B through a network cable under optimal conditions. Everything else is up for negotiation, whether it's an on-premise or cloud-based solution.

It is my experience, though, that a decent amount of critical thinking is required when you dip your toes into your first automation project. The novice in both management and applied craftmanship may cry, "Automate everything!" without giving abstractions such as opportunity cost and investment up front much thought.

Automation doesn't come for free—far from it. A higher level of automation is likely to increase the complexity of your organization and technical landscape. You may also in different ways be pushed into acquiring new skills and knowledge and—yikes—end up working differently. Roles and responsibilities are prone to change when automation projects get priority. The cost embedded in motivating, training, and getting people to work together in new ways is an often overlooked and hence hidden cost. Inadequate change management is a risk factor that you and your organization will need to attend to. What will the benefit be from buying the latest, hottest tools and advocating the right principles if you as a leader are unable to get your people to use them or they don't feel inside that it's the right path you're leading them down?

Automation is, in layman's terms, letting a machine do work that was previously done by a human being. It takes willpower and patience to make the right choice at the right time when people's routines are affected—which by definition will be the consequence of letting a machine or a series of zeros and ones take over on assignments that were manually executed by a human being the day before. You have to realize that from the very beginning, an automation of workflows will change your way of communicating. When manual, repetitive labor transforms from human workflows into workflows configured in a software solution, it could also potentially change the expectations that your leader will have for you. That applies to the developer or IT technician who needs to transform concrete, manual workflows into series of zeros and ones, but also to team leaders with HR responsibilities. Leaders tend to have a leader as well, a fact that people on the floor without HR responsibilities in my experience tend to forget from time to time.

Computers stand without competition when it comes to repeating the same task again and again, tens of thousands of times without interruptions, without complaints, and without the slightest variation in the quality of work being done. Humans are quite the opposite. We're notoriously bad at routine labor. The variance is simply overwhelming, even when we focus every cell in our body and concentrate on producing repeatable results. Try, for instance, to draw twenty exactly round circles on a piece of paper. Not two will be alike. They may look alike at first, but they aren't if you zoom in a bit. A computer attached to a printer would be able to print twenty identical, perfectly round circles. It could print a million if you were standing by feeding paper into the printer and replacing toner cartridges when needed. A computer can produce the same output again and again and again because that's what machines and computers do. In the middle ages, books were manually copied by monks from one book to a new one with blank sheets, until Gutenberg in the mid-1400s invented the printing press and automated that process. What we do with software and automation today in the twenty-first century is essentially the same process; the context is just very different from Gutenberg's.

Machines and automation remove the analog imprecision in a human being. What human beings can and do that machines cannot is exploit the powers of our brain to explore possibilities and get ideas no one has ever had before. We people can choose to break the rules. A computer never does that unless we program that possibility into it. We humans are capable of abstract thinking. We have self-awareness. We develop and learn new skills. We think "what if?" and get crazy ideas. We are the sum of our ancestors' legacies and personal experiences, and we have developed adaptable skills that allow for incomprehensible flexibility, even though the human being's physiological expression and genetic landscape hasn't really changed during the last 10,000 years. It is impossible to understand how many abstractions you must be able to comprehend if you compare living and dying 10,000 years ago in a Stone Age village, never interfacing with more than perhaps fifty people in total throughout your lifetime, with living and dying in a modern city like New York or London. Yet we have accomplished that journey as human beings, slowly adapting and adjusting to new ways of fulfilling everlasting needs for food, shelter, reproduction, and safety.

Mankind has invented and developed computers and related technologies to assist us in our continuous hunt for efficiency. What people and organizations still have trouble with is understanding that computers and technology are more than just a laptop, a suite of Office products, and a stable internet connection. Technologies and software are the engine liberating human beings to work at what humans do best: making fast decisions based on experience and with little data, being creative and figuring out visions and openings no one has seen before, and spotting a market even before the market knew there was a need.

As a prerequisite for success on the automation initiative that I imagine you and your team will engage in shortly depends establishing a foundation of trust with your leadership team. They need to know that the company will eventually earn money on the initiative that you are advocating.

Be aware that you may find a somewhat reluctant or maybe even hostile attitude from your potential sponsors and stakeholders toward the notion that automation is important. All they see is more expenses for licenses and custom programming of something abstract. They don't understand that IT and automation have the potential to give the company a competitive advantage due to the increased and well-documented quality in produced goods. You know all this, but you shouldn't assume that your leadership team has these insights or is ready to absorb them. You can't assume that your leader understands the notion of computers as a machine that is capable of producing output without variance in ways that human beings cannot.

It's likely that you and your team will have to start from ground zero and prove that the project is capable of becoming a cash cow. It won't be enough to lay out a roadmap that implements a strategy with an unspoken value proposition because you cannot fathom anybody will question it. You know that automation, when implemented the right way, increases bottom-line revenue through operational cost savings and at the same time increases employee job satisfaction. It is evident to you that the company should engage in initiatives that motivate the highly skilled (and highly paid) knowledge workers that the company depends heavily upon. The worst thing that could happen would be that these employees would leave the company for marginally higher pay and better cafeteria arrangements at a competitor, right? Don't assume that good people above you in the company realize that. Their perceptions of cause and effect on a topic may differ greatly from yours.

"Okay then," you may ask, "What do we do? How do we get started so the leadership gets to understand the opportunities and value proposition in what my team and I want to do?" The answer is that in reality, only you, your colleagues, and your management can provide you with an truthful answer about what you should aim for and how much you might invest in an automation initiative. I can help you with counseling and guidance in the forthcoming chapters, but you are the only ones who know what makes sense in your context.

As I mentioned before, everything can be automated, otherwise Google, Netflix, Amazon, and the rest of the big boys in the schoolyard wouldn't have become the companies they are today. Bear in mind that just because these industry-defining companies are world champions in automation doesn't mean that you should pursue the same strategy as they do. It's perfectly okay to copy methods and processes from experts who have written and delivered keynotes at conferences. Copying others isn't a flawed strategy at all.

If you're starting at rock bottom, the strategy shouldn't be automating everything. You'll die trying, is my bet. Instead, build a foundation of quick wins that allows for motivation to progress. Then keep investing in larger, more complex initiatives, focusing on applying leverage on earlier phases of automation wins or failures in your organization. Start by doing what others in your situation have had documented success with before. Base your first initiatives on standard, off-the-shelf software that you can throw away without too much hassle. If it turns out to be a bad fit for your organization, replace it with a new candidate.

Be aware of cargo-culting[1] and Not Invented Here (NIH) syndrome[2] because you can potentially pour a pile of money into a tarpit by engaging

---

1 "Cargo Cult," Wikipedia, last edit June 19, 2020. https://en.wikipedia.org/wiki/Cargo_cult
2 "Not Invented Here," Wikipedia, last edit June 20, 2020. https://en.wikipedia.org/wiki/Not_invented_here

in initiatives that are supposed to automate complex workflows and processes if you don't build feedback loops into your initiatives in parallel with the execution of your automation initiative. If you start out in the wrong direction, you may very well end up with missed opportunities and low return on investment (ROI). You might have gotten much better feedback and achieved happier end users for a fraction of the cost if you'd prioritized other initiatives besides the one you pushed for without doing your homework and building a business case up front.

Build a Business Case for the Advantages of Automation

Changes happen when we decide to act differently. Studies in behavioral design indicate that lasting changes in behavior rest on a foundation of inner motivation.[3] It is important before trying to get buy-in for a project that requires change, such as the implementation of an automation initiative, that you build a shared language and common ground. You should invest in building your case so that key stakeholders feel some sort of inner motivation to actively support the change in concrete behavior that is necessary for the automation initiative to be a success. If we assume that this initiative is best understood by you and your team, it is also your responsibility to do your best to transform your knowledge into a language that stakeholders can understand if you want to ensure long-term success and future investments in automation.

I recommend *Find Your Why* by Simon Sinek if you want to read a bestseller on how become proficient in communicating why what you say should make sense to others. Sinek has made a living out of inspiring other people to inspire others. The book has sold millions of copies, which just underpins how complex it can be to figure out what motivates people.

Sinek has codified his experiences from his life as project manager and CEO into a simple framework he calls "The Golden Circle." It describes how you should have a foundational principle or a set of values that serve as your guiding star. This guiding star should be clearly visible in everything you say

---

3   John P. Kotter, *Leading Change* (Boston: Harvard Business Review Press, 2012).

and do, on every channel, toward every stakeholder, customer, or lead that you will ever encounter in your line of duty. It all comes down to the fact, according to Sinek, that people don't buy into *what* you do—they buy into *why* you do it. It is your ability to communicate your why that determines if people will engage with you and decide that you're trustworthy and have credibility. If people are convinced and believe in the same values that you do, they're more likely to buy your products

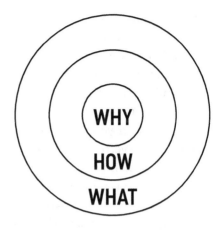

Figure 1. The Golden Circle, Simon Sinek.

regardless of what they are. The products are merely an expression of your shared dream of belief. I recommend, too, watching Sinek's TED Talk video.[4]

I'll be the first to admit that regardless of how many pages I write about change management and how to do—or not do—stuff, it's likely that you'll need more than just this book to cook up a plan for change initiatives. I urge you to attend meetups, read more books, talk with people smarter than you, and ask how you should plan for building a business case.

Building Better Software will provide you with a framework for asking questions that may inspire stakeholders and colleagues to reflect. It will, I hope, ignite interesting discussions that will provide data for your shortlist of ideas for an implementation roadmap. You will work with others as well as using your own motivation for wanting to go in a different direction. Other human beings get up and go to work every day with the intention of doing their absolute best for the greater good, but their measurement for being a good corporate citizen may very well be in direct conflict with the business case you're trying to build.

It's essential for you to figure out why automation constitutes a good thing in your company because Continuous Integration and Delivery in the forms of build and deployment pipelines aren't your why, per se. It'll be the

4   Simon Sinek, "How Great Leaders Inspire Action," TEDxPuget Sound, September 2009. https://www.ted.com/talks/simon_sinek_how_great_leaders_inspire_action

natural evolution of a more profound understanding of cause and effect related to knowledge work and how business value is being created in the twenty-first century. You'll never convince anybody of the advantages of automation if you are unable to explain to them clearly what problem automation addresses and provides a solution to.

If you put "Continuous Integration" in the center of your Golden Circle, you'll fail. But if you put some words in there that express "world-class quality," you're getting closer to values and beliefs that people like yourself, regardless of their place in the organizational hierarchy, want to be a part of. Who doesn't want to deliver high quality when they go to work? If you can (and this is the hard part indeed), you should be specific about what you really want to achieve then build your case around why automation is the right answer. Now "automation" will not be in the center, but it will be in either the "how" or "what" circles, depending on the way you frame it. It suddenly becomes a concept that bridges what you believe to potential solutions and actual execution of achieving the world-class quality that everybody wants to be able to deliver. The road to hearts and minds doesn't appear to be that long after all. It's just a matter of starting out right in your communication efforts.

The example from above is nothing more than just an example. It's not a plug-and-play framework that will work everywhere. The reason that I'm spending a bit of time on Simon Sinek and the Golden Circle is that I often see very skilled employees in companies who experience great difficulties in converting their domain expertise and skillset into the forums where decisions of strategic importance are being made. Rather than blurting out, "Well, it just does!" when asked by your leader why automation makes more ROI than prioritizing the top two or three customer features requested, you may say instead, "We don't know as a team that we're capable of delivering an acceptable level of operational stability unless we remove the human factor from our deployment strategy." Now you've framed the discussion very differently, allowing both sides to engage in a more constructive dialogue.

Your stakeholders will need to understand why an automation initiative is an advantage to them and everybody else. You and your team should work hard to avoid making the same mistake as countless software teams have made before you: not taking the human factor into account. Never forget that any project or action where people will be required to work differently together is in part also a change management project as well as being about building concrete, shippable software deliveries. Automation will streamline the surroundings of each individual contributing, meaning that each individual in an organization, from top to bottom, will be forced to face different expectations when processes and work is automated. It isn't a goal in itself to change context necessarily, but it will happen eventually when you engage in an automation initiative.

You and your team should invest time and willpower into writing down what you really want to achieve. This process allows for you to think for a while, take a break, and then rethink your strategy based on a discussion of business value, organizational impact, and expected ROI. It's often easier to design a better manual process than to automate a flawed one if you're unsure what a good process looks like in your context at all. It goes without saying that if you have a messy process and you automate it without redesigning it, all you will get is an automated mess. That rarely benefits the company and its shareholders.

You can use a simple technique such as the Five Whys[5] or the Golden Circle in your analysis of your stakeholder landscape. Plan your communication efforts so that they can see why an automation initiative will benefit them given their position in the office hierarchy and job description. Let's imagine for instance a tech support figure; we'll call him Greg. He lives and breathes to deliver good service to those calling in or creating support tickets. He's a genuine good guy and a great colleague. Everybody is impressed that he is still able to maintain a positive attitude and smile even at three in the afternoon when the phone is flashing for the thirtieth time that day with yet another disgruntled office employee. His version of "the best I can do when I go to work" is to smile through the phone, document decisions and department meeting summaries, and advise other departments when new software is about to be released. There is a set of distinct tasks that he is responsible for, such as sending out release notes and performing a series of manual ad hoc tests upon requests from the software delivery teams after software has been released to end users. He is absolutely crucial for any company—helpful, timely and good at what he does.

---

5   "Five Whys," Wikipedia, last edit June 29, 2020. https://en.wikipedia.org/wiki/Five_whys

But what are the odds are that Greg will welcome an automation initiative? It will greatly impact his daily list of chores and tasks, and potentially endanger his job position by making it obsolete. There won't be as many messages to send out; a software solution will take over going forward. The tests that he executes manually might also be executed by a piece of software that also send out a report without any human intervention from his side. There may still be a need for publishing release notes, but the notifications about software being released are also likely to be programmed into a software solution doing software deployments and release management.

Greg is an important stakeholder that you'll need to handle and spend some time with when you decide to execute automation initiatives that can completely overturn the work and the expectations that meet him when he walks through the office door every morning. Even though he may not have much formal influence, you must never underestimate the importance of getting someone like Greg on board with your automation initiative, especially if he has a lot of trust among and relationships with his peers. If by doing your stakeholder management early on you can convince Greg to become an ambassador rather than opposition toward your automation initiative, your position will without doubt be strengthened. He'll be helpful when the automation is about to be implemented on the technical level but also inside the organization, where lots of other stakeholders might be affected just like Greg. You will need ambassadors in your organization, and you will need to prepare and include them well in advance in order to succeed later on.

Another example of a stakeholder could be the leader of IT operations; let's call her Sarah. She feels a loss of control when manual labor is suddenly taken over by software with workflows that invoke changes without any human intervention whatsoever. She may use passive or even active resistance against an automation initiative because development teams in other parts of the organization besides hers are allowed to deploy and perform changes on "her" infrastructure. She's actually comfortable with the fact that development teams need her team to manually push deployment buttons. The sense of control and ensuring an acceptable level of change auditing are fundamental values to her. Auditing changes may be what Sarah feels is expected of her when she goes to work every morning.

I've included Sarah and Greg to show how important it is to be curious about your stakeholders and engage with them early on. You may have stakeholders in your organization who, for reasons of their own, resist your change initiative, even though rational arguments of the potential for improvements in the company's ability to deliver high-quality software with predictable costs. Software developers don't always realize that they are often the only ones who know the entire value chain from A to Z. They are in reality the only ones who are able to truly optimize anything without suboptimizing single steps.

This is why some companies, again and again, across projects and initiatives, are completely dependent on very specific key resources in their IT department, sometimes down to one or two people. These are the people who understand the entire chain of events, from a customer entering the website, putting something in the basket, paying by credit card, and creating an order, which is transferred into the ERP system. So far, so good. But when the product is being returned because there was damage during transport, a whole new chain of events is initiated. The developer also knows every intimate detail of this process because she has been part of assembling and configuring all software systems handling all business workflows throughout the entire value chain—not just the one describing how this business wants to provide a solution for ordering and delivering stock from the main warehouse.

Greg doesn't have this knowledge and the deeper understanding of the process; nor does Sarah and her team, who are responsible for operating servers running the various software solutions across the organization. In the worst case, Greg and Sarah may see you and your team as an element of distraction coming along with a set of initiatives that only introduce friction and unnecessary noise.

Both sides may have their reasons for thinking the worst of the others. You just have to ask yourselves, if you're a software development team reading this, Who is best prepared to take initiative here to listen to what others have to say, the ones who know something the others don't know or the ones who may not know what they don't know yet?

It's you, of course, who will need to come forward and communicate how you see an opportunity to optimize the entire value chain to the benefit of all of you. If you're in doubt about how to make a plan of attack for doing so, the following exercises will help you get started.

Two Exercises to Help You Get Started

It is pointless to set the expectation that if you go through exercise A, B, and C and receive a score higher than sixty-five, you're ready to begin planning and executing your upcoming change initiative. Machines work like that, but not human beings. You need other strategies and other tools to influence the stakeholders that you and your team work with as part of your daily jobs. Prepare yourselves for the fact that you and your team are likely to keep working together with your stakeholders, probably over the course of several meetings where you follow up and adjust according to a master plan or vision for what you want to achieve.

Below you'll find exercises for pulling in stakeholders as needed to build relationships, explore historical reasons for current ways of working, and build a shared language around shared pain points. You may even find a few allies while doing so that you didn't know existed.

The first exercise is called "What's In It for Me?" This is an extension of Simon Sinek's Golden Circle. It will help you map out every stakeholder's motivations for and against an initiative that will in one way or another affect their daily lives.

The second exercise, which I for the lack of a better name tend to call "Error and Incident Analysis," focuses on your end user's perspective of the quality of your solutions. Looking from the outside in, we will try to establish the motivations for concrete actions or misalignment in quality as it was experienced by your end users. The data foundation will be the incidents and outages reported by your customers and end users that have been fixed subsequently by your organization.

These exercises do not need to develop into workshops. Keep it simple, and if possible use whatever collaboration framework that you are using already. It could be a topic for a monthly department meeting, as inspiration in a one on one between team members and their line manager, or just something you chat about during breaks by the coffee machine. It's okay to focus for a few hours going in-depth with cause and effect; one doesn't exclude the other at all. I believe with all my heart that it is the small, repetitive conversations and nudges that create change in the long run. The large, staged off-site where the entire department or organization goes for an all-inclusive weekend trip to all talk values and missions, and the senior leadership team reveals its new strategy that was devised during a leadership retreat last month, rarely makes a difference. It just doesn't work.

It is at home, in your daily work, that you as a team need to enforce new agreements about changes in actual behavior. That requires leadership skills and focus, and it starts out small. Don't overcomplicate things. It's much better to have three or four thirty-minute meetings where you follow up on previous actions and repeat decisions and visions rather than a 120-minute meeting without any time to do follow-up afterward.

....

What's In It for Me (WIIFM)?

Before proceeding to the work ahead of you, you will need to relate to the following two sets of questions:

1. How will (insert stakeholder group or name) benefit from having automated software deployments?
2. Which concerns could (insert stakeholder group or name) have related to automated software deployments?

Here are a few examples:

- "How will our users in Germany benefit from having automated software deployments?"
- "How will our leadership team benefit from having automated software deployments?"
- "How will Debbie from marketing benefit from having automated software deployments?"
- "What concerns could Debbie from marketing have related to automated software deployments?"

Find more in your own context. It's a simple formula, which is why it can quickly open up a series of rather interesting discussions in practice. If, let's say, a DevOps initiative focusing on automating workflows and software delivery has been decided top-down, you should want to include leadership teams and sponsors if you haven't had a discussion with them about their reasons and motivations for doing so. What do they really want to achieve on your behalf? You may want to invest a bit of effort into that dialogue up front so you know you're aligned going forward. You should definitely be able to answer questions like the ones above before proceeding with your automation initiative.

Question techniques such as the Five Whys can help teams see the world from another team's perspective and, given their situation, describe what they will gain. Include and explore the limitations and obstacles that a stakeholder like Debbie from marketing sees coming toward her in a future where parts of her job description may be taken over by a piece of software.

Can you turn a change or a reduction in responsibilities or work assignments into an opportunity to take on more work of interest to a stakeholder? What possibilities may emerge for you as a team leader by automating manual processes in your team? Our friend from earlier, Greg the supporter who may foresee that he is going to be out of a job, may be a good fit for taking on new assignments that nobody had time to do before because every available timeslot was spent doing manual labor. Maybe he really wants to do more focused regression testing leveraging his deep domain expertise in the systems that end users are facing every day. Maybe he would be able to sanity check business requirements early in the process by drawing on his qualitative insights into end user behavior to ensure that they are aligned with processes as they are actually implemented in the business. I know for a fact that having somebody like Greg available to you on a software development team while uncovering functional as well as nonfunctional requirements and assessing risks of various proposed solutions is a competitive advantage. Whenever change happens, doors close, but if you look for them, you will find doors opening too.

You and your team could come up with your interpretations for why and how your initiatives may benefit the greater good, but if you have the opportunity, reach out and ask Sarah or Greg. If you can't just walk by their desk and ask them, use the fact that most people are dependent on their Outlook calendar to your advantage. Book them for a short fifteen-minute coffee break and tell them in advance that you're curious about knowledge that he

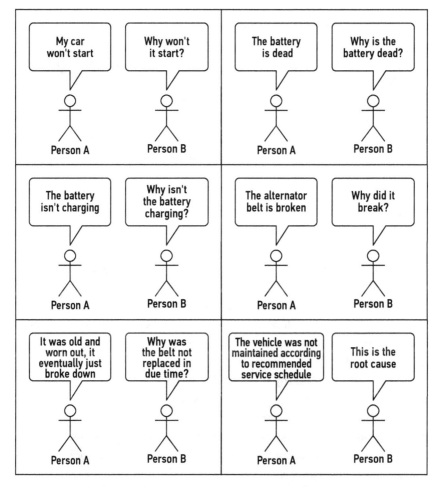

Figure 2. Example of a Five Whys session investigating root cause of a car not starting.

or she possesses. You'd like to pick their brains about their reflections on the upcoming automation initiative. If you have established a sufficient level of trust, you may also ask about inner motivations and how a great day at work looks to them. This will extend your own understanding of your stakeholder's concerns or worries.

Reach out; it doesn't cost you anything. On the contrary, the chance of positive feedback is overwhelming. People want to be asked and included as subject matter experts in their field (watch out for the occasional foghorn, though). No human being with a bit of logical sense will question you for reaching out and investigating how change initiatives will affect the organization around you. Take that first step, it's okay.

The purpose of doing this exercise is to get acquainted with your stakeholders and understand the organization from their perspective. With this chapter in mind, you ought to at least try to understand the concerns and troubles that stakeholders may have prior to upcoming changes in processes and expectation alignment as it affects their job description. The better you're able to understand and avoid judging stakeholders as old-fashioned or thickheaded bureaucrats just because they voice reasonable concerns toward the effects of an automation initiative, the better off you will be in terms of tailoring a roadmap that sufficiently adapts to everybody's need for speed, transparency, and quality in deployment of software deliveries.

You're not going to solve their problems immediately; that's not the case with these exercises. You should express that clearly if stakeholders expect you to solve their issues. Focus at this stage should be on learning and relationship building. You will need both later on in the process. Build alliances and sharpen your communication skills along the way; it's likely that you might learn a thing or two about how things work in your organization. If you manage to build healthy relationships with stakeholders, it would be my advice to nurture them going forward, maybe with small, biweekly check-in meetings, maybe for a few months. Cake and lunch meetups tend to work out really well, too, if you see an opening and can meet in person.

End users are, for reasons beyond most of us, often overlooked stakeholders in some organizations. What would the end user get out of all of this? The next proposal for an analysis might help uncover just that.

Perform an Analysis of Errors and Incidents

If you and your team are working on a piece of software that has been in production for a few years and you have a handful of end users already using it, there is likely to be a list somewhere of errors and incidents that you have diagnosed and fixed in your software. These data are invaluable when you need to understand end user behavior in your systems or their perception of the level of quality you are delivering to them. I would expect that you have a list of errors and incidents in a ticket system somewhere. It may just be a spreadsheet on a shared network, which is fine, or it may be an enterprise ticketing system. The form isn't very important. What you need to do is to extract a list of all incidents going back six months or so to find the documentation for all the errors and incidents reported, diagnosed, and fixed.

Data about each reported item should be placed into the following matrix in the green fields:

Incident ID	Title	Reported by	Incident description	Why did it happen?	Why?	Why?	Why?	Why?

Figure 3. This matrix is heavily inspired by the Five Whys.

If you need more data, such as a URL deep linking to the ticket, feel free to expand the matrix. Once you have extracted necessary data from your ticket system, you should walk through every single incident and use the Five Whys to uncover the root causes of each incident.

Below is an example of raw data filled into a row in the spreadsheet above:

Incident ID	Title	Reported by	Incident description	Why did it happen?
4711	Facebook login doesn't work	Debbie from Marketing	When creating a new customer and clicking "Use Facebook", the page goes blank and eventually returns a technical error message. We're loosing a lot of customers on this!	

Figure 4. Start your investigation by gathering data. The most important thing is to describe how the system doesn't live up to expectations in each case. You may want to add extra fields to hold more information. This is just an example.

Now you can investigate the root cause using the Five Whys. The fictional conversation below shows an example of how to uncover root cause.

Q: What was the error?
A: I opened the website and got an error trying to log in with my Facebook account.

Q: Why did it return an error? (first Why)
A: The Facebook login hasn't worked since the release on February 1.

Q: Why not? (second Why)
A: The API key for our Facebook login was incorrect.

Q: Why was the API key for the Facebook login incorrect? (third Why)
A: One configuration setting was incorrect on LIVE, meaning that Facebook rejected our login requests in their API.

Q: Okay, how could that happen?
A: The wrong configuration file was copied from the developer's workstation without anybody noticing that it was the wrong file.

Q: Why didn't anybody notice? (fourth Why)
A: We don't test before a release that the configuration files are the right ones for our environment.

Q: Why don't you perform regression tests prior to a release? (fifth Why)
A: There is no process in place to ensure that only valid configuration files are copied into production.

It sounds easy, but I can tell you from experience that Five Whys sessions can be exhausting in real life. It's not always true, but focusing on finding and learning from previous mistakes can be difficult and painful to participate in. It may be difficult because you shift focus from fixing an error to finding the root cause of the error appearing in the first place, so you need to actively consider every detail in the chain of events leading up to the end user filing an incident for your team. That level of detail may be hard to uncover, and you may end up in situations where nobody really knows what went ahead of a certain incident because it is impossible to deconstruct the chain of events in the past leading up to something like an outage. You and your team are forced to inspect the nature and consequences of concrete behavior and actions taken—or not taken—in your team in the context of end users experiencing outages or low-quality deliveries from your team.

In the example from above, it may turn out that you actually did have a process for testing configuration files, but it went a little too darn fast that day, and it was late, so you skipped the very test that would have warned you that you were about to introduce a login error. Maybe it was the developer with the highest salary on the team who also has a hard time admitting when he's wrong. His knee-jerk reaction is probably to take on an aggressive attitude defending his actions. I figure you've come across that specific personality type at some point in your career already.

Maintain focus on the job at hand. This is a learning experience, not a finger-pointing exercise. It is indeed possible that you will need an independent facilitator who can coach your team and take the lead during root-cause analysis of incidents and outages. It's hard work—been there, done that—but you shouldn't shy away from doing Five Whys or similar exercise to uncover cause and effect behind an incident. Think of all the opportunities and learnings that you can extract! You may very well get to know what goes on inside the bowels of your organization and how processes are actually lived out—processes that contradict earlier agreements made during the staged, all-inclusive off-site where everybody participated in dining and shoulder-patting and talked values and agreed that you would go with a four-eye principle for all changes on LIVE going forward. No more having developers putting changes on LIVE alone, risking taking down the entire website. No more looking like a bunch of amateurs because marketing was pressured by senior leaders to release a campaign that wasn't yet copyedited. No more team leaders who cannot delegate responsibility downward and let the team take decisions on their own. I figure you know the drill.

This is why a list of errors and outages is such a valuable container of data. It is a defined data source directly connected with your end users and with a high level of trustworthiness. It showcases the low quality you and your stakeholders wish to improve. Every single reported incident should be considered the beginning of a conversation with your stakeholders about expectations towards the feature set and the quality of those features that you're releasing to your end users. Pick up this great opportunity lying right before you—your stakeholders and end users will love you for doing so when you help them to understand why you're reaching out. If this is the first time they hear back from you, you should prepare yourselves to be yelled. You could also write up a short statement about why you couldn't reach out months ago. You get what you deserve, I guess. This is still the best opportunity to engage with your directly involved end users. Better late than never.

So take all errors and outages being reported that you and your team have fixed for a period back in time. My personal rule of thumb is six months, but if three or twelve months works better for you, feel free to adapt.

For every single outage, you should take a deep dive to uncover the root cause. Avoid coming up with new solutions to an already solved problem. Focus on revealing hidden knowledge and collaborate around realizing shortcomings or the conditions of your codebase, your processes, or your abilities to pass on information to other stakeholders. Be aware that there isn't magic involved in this. A root cause doesn't necessarily materialize just because you manage to ask "Why?" five times in a row. Sometimes three is enough, sometimes you need six or seven takes.

I usually stop looking when a "Why" is answered or shows evidence that a human being in a certain situation made a choice that led to the incident in question. This is the behavior or choice that you don't want to repeat over and over again. It may be that a choice was made due to incomplete or flawed data from, for example, a monitoring system that reported everything was okay after a release without any systems responding to end user requests. Very well, stop there. Now you know that there is work to be done recovering trust toward your monitoring solution. Perhaps you need to adjust your process of validating a release by introducing extra steps manually until the monitoring system is working again to the extent that you can trust it when it reports everything is okay in the future. Don't forget to discard the temporary steps once monitoring works again.

If you have a heap of incidents that have been fixed because the nature of your applications or landscape simply breeds lots and lots of incidents all the time, I would focus on certain areas of interest and leave the remaining parts untouched for now. There is a risk that you may look in the wrong

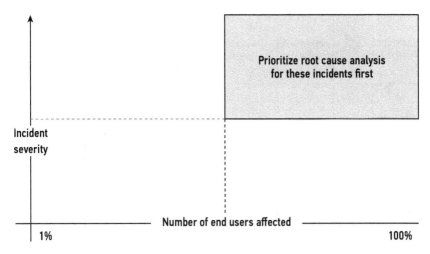

Figure 5. One way of filtering large datasets of incidents is to map all your incidents for a given period of time into this matrix as a group exercise. Then focus your root cause analysis efforts afterwards on the ones with high impact of the most users first (e.g., the ones you've placed in the top-right square).

direction for learnings and improvements and end up suboptimizing your processes, but there's no reason to do root-cause analysis on hundreds or perhaps thousands of incidents. You can't embrace everything, so you'll need to start out small and focus in one direction. Perhaps reiterate a few times in different directions over time.

If you decide to start by filtering the list of reported incidents prior to doing root-cause analysis, focus relentlessly on errors with high end user impact for a majority of your end users. I would suggest that an error to login functionality by definition has a high end user impact. Other examples may be printing out invoices, adding items to a shopping basket, or similar. It may be less important that an export function to CSV files didn't work for a period of time, even though this error will affect end user experience. But if only one single user experienced the error and he could work around this error until you provided a fix, it isn't as important to uncover root causes here relative to investigating root causes for introducing errors to your Facebook login. Maintain a focus on where your efforts in terms of increasing quality would matter the most to a majority of your end users, and then work your way through the errors and incidents reported in those areas.

Incident ID	Title	Reported by	Incident description	Why did it happen?	Why?	Why?	Why?	Why?
....	....	....	....	....	....	....	....	....
....	....	....	....	....	....	....	....	....
....	....	....	....	....	....	....	....	....
....	....	....	....	....	....	....	....	....

Figure 6. Once you have gathered data, expand your list with "Why" columns to capture cause and effect leading up to the incident.

Once you've completed this exercise, you should have a list like the one in Figure 6.

You may stop here and take the lessons as they are right now. You've probably learned a great deal by now. It's perfectly fine to add a few extra columns as needed if you need to hold on to some extra information along the way.

What you have produced now is a series of lessons learned about your organization, about your processes, and about the state of your software and enterprise architecture. Those learnings will be key points of reference when you build your business case, which should document how your business will benefit and use a change initiative. You will add a lot of trustworthiness when you state your case if you can go back to a list of documented shortcomings and explain why it's necessary to adjust behavior, change processes, and perhaps speed things up by automating away variance and unnecessary friction caused by misalignment, lack of documentation, waste incurred by improper communication between departments, or something completely different—whatever you uncovered during your root-cause analysis.

You shouldn't look for ways to automate every single process that has flaws in it. It's perfectly sound to suggest an improved but still manual process, especially for tasks that happen rarely and can be planned for in advance by teams. Write a better and improved wiki-article with screen-shots and perhaps a small video walking new employees through what they should do step by step. That only takes a few hours and may be a much wiser choice than spending hundreds of hours automating the process all together.

Look to automate the small, tedious tasks that you do all the time. The greater the volume, the greater the benefit of automation—that's the rule of thumb you and your team should be going for.

....

Summary

There are many ways of conducting root-cause analysis, post mortems, and similar retrospectives of past performance. Common to them all is not judging individuals for their actions alone. If somebody didn't fol-low the process, take it for what it is without judging the person as being incompetent or a bad person. You don't have to agree with me on this, but I still believe that everybody goes to work every day trying to be the best corporate citizen they can be and doing their best every day. It's with open eyes that I decide to think this way about the motives of others because it helps me remain curious as to why a business associate or colleague didn't pay attention or didn't follow direct orders. One doesn't place files in the wrong folder on purpose when you want to be a good corporate citizen every day, so why did it happen anyway? Lack of communication? Flawed assumptions? Ambiguous documentation?

It's very common for people to be reluctant to answer to questionnaires or feel they are being interrogated if you and your team show up and ask lots of questions about things that didn't go wrong. People won't give you what you need if they don't trust you, so refrain from formalizing this data-gathering exercise unnecessarily. You're likely to get nowhere fast if you only ever see a stakeholder while sitting in a meeting room. Ask Sarah from IT operations if you can meet her for a cup of coffee instead, or if you can chat with her for five minutes after your next department meeting if she holds valuable information or might be able to tell you more about the historical reasons for having a process of X. Keep it simple and vary your ways of approaching people.

When you conduct your analysis and ask around, you should be aware that pathological, risk-averse organizational cultures might decide that you're a rebel trying to bend or break the rules. Keep focused and don't judge. I agree with you: it's easy to say, hard to do, and you're likely to stumble a few times along the way.

Part 2

Document Your Existing Release Process

IN MY EXPERIENCE, lots of automation efforts are wasted in organizations because teams and individuals do not spend time documenting current processes and workflows before beginning to improve and automate their processes. It is vital to understand your current processes for building and deploying your digital deliveries; otherwise, you may end up automating a different process than the manual one that's actually being performed in your organization.

You and your team should document every single activity in the series of events that in total encompasses your entire release process as it is today. You will need to be able to explain to others what you do, if and when you need other departments to help you prioritize things such as service requests that the success of your automation initiative depends on. Don't worry—documenting current ways of working isn't difficult or complex stuff. It's very concrete and visual, and I'll give you a recipe to help. The outcome will be a collection of shared knowledge, in the form of diagrams, that you can show to others when you turn to your stakeholders for support and investment in the initiatives you are going to propose.

In the following chapters, I will show you concrete exercises that will help you to document and understand how the processes used to build and deploy digital artifacts take place in your context today. I'm assuming that you're on a team, but the early stages of the work could also be executed by a single individual.

What you should ensure is that the diagrams you create only describe knowledge you've confirmed. You should know for sure, not assume, how things occur in your context when you document your processes in the following chapters. Faith in any documentation deteriorates if you and your team make assumptions about the nature of execution of tasks. It's crucial that statements from others are validated by cross-referencing those statements with other stakeholders. Include as many stakeholders as possible, to the extent that it makes sense of course. Create a list of unresolved issues in a parking lot for subjects or statements that you want to verify. You can either ask others for qualitative input or gather quantitative data. This information may or may not add credibility from an independent source regarding statements made during a lively discussion about your current ways of working.

What's a Parking Lot?

A Parking Lot is essentially a bullet list of unresolved items or insights that may spin off from discussions or one of those great ideas that suddenly comes into mind but where timing or context doesn't allow you to investigate further.

If you get one of these "aha" moments while collaborating with others while using whiteboards or similar, you should designate a square on your wall or whiteboard and label it as your "Parking Lot." Put sticky notes for every great idea or follow-up task, such as "Remember to update vacation calendar" or "Did vendor X ever get back to us regarding Project X?"

Parking Lots allow everybody to use limited human bandwidth to focus on the job at hand without forgetting important insights or action items.

I'll help you and your team to explain and visualize the entire chain of necessary events, from changing a line of source code on a developer's workstation until this change has been promoted to the systems fielding traffic from your end users. You and your team need to take accountability for knowing all the activities in the entire chain of events—not just parts of them. If there are large areas of black-box magic happening, you will only optimize what you already know, not the entirety of the value chain. You risk suboptimizing your combined process, even though single activities may perform better out of context. If you do not have a holistic image of all activities in the chain, from deploying a change of source code to production, you and your team will not be able to optimize based on an understanding of how changes in one part of the value chain affect activities further down the value stream. It's a classic antipattern in LEAN to optimize based on incomplete knowledge and understanding of the entire chain of events.[6]

To mitigate the risk of working without having a bird's-eye view, it's essential that you start by looking at what you're doing in terms of concrete, tangible actions today. Practice communicating that knowledge so others can verify your achievements before you and your organization decide upon a new direction. The chances of the success of your automation initiative depend heavily upon your abilities to take each stakeholder's domain knowledge into account when you assess your current processes. It is in collaboration with stakeholders who are performing existing processes that you can pinpoint and formulate the crucial short- and long-term benefits of automating manual processes. I recommend that you argue and push quite persistently to obtain the necessary system-wide understanding of all processes to ensure that you comprehend your current situation before trying to improve it.

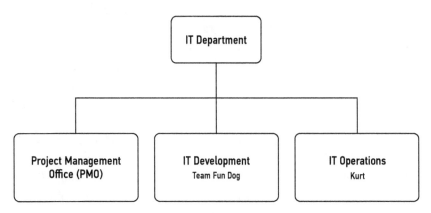

Figure 7. Organizational diagram with IT Development and IT Operations in separate legs in the organizational hierarchy.

---

6 "Suboptimization," Merriam-Webster.com. https://www.merriam-webster.com/dictionary/suboptimization

Prerequisites

In the examples to come I'll use a team of software developers who call themselves "Team Fun Dog." They have a production-like environment called "TEST," which only they and other employees in their organization have access to. There's also a production environment that I call "LIVE," which is the one responding to and serving end-user requests from the systems that Team Fun Dog are releasing minor and major upgrades for.

There are two distinct roles described in the following examples: a software developer role at Team Fun Dog, and a database administrator I'll name Henry who works in IT operations (see Figure 7). In real life there will be many more roles involved end-to-end in delivering software to end users, but for the sake of this example, we'll just work with these two.

I've decided that both Team Fun Dog and Henry in IT operations are working in a classic IT organization where software development is handled in one department of the organization and IT operations takes place in another department. IT operations takes care of operations and supports solutions deployed and released by IT development. There are many variations to this pattern of segregation of duties in companies, but lots of companies orient themselves in a similar way. There are some companies that retrofit the organization to improve support of full-featured, cross-functional teams.[7] The point of having Henry and Team Fun Dog in each of their departments, with segregated duties and job descriptions, is that it creates an organizational complexity. I will refer back to this example as we work our way through the exercises and discussions throughout the book.

Team Fun Dog is working on a web-based product, which is why there will be terms such as "webserver," "database server," etc. in example diagrams, but they could in principle be working with just about any digital delivery. Video production, embedded software, Internet of Things technologies, or an app for a smartphone would fit just as nicely as a means of a delivery in the context of deploying digital deliveries to end users. Team Fun Dog, in

---

7 "Future Teams," LeSS, accessed July 15, 2020. https://less.works/less/structure/feature-teams.html

my example, is working on a web-based product since most software developers and IT operational staff have a high-level, conceptual understanding of web-based technologies and the protocols involved.

> I've embedded a bit of organizational complexity on purpose in the examples that I'm using. Lots of teams and skilled IT technicians will likely nod in agreement when I say that organizational inertia and conflicting priorities across organizational boundaries can be difficult to overcome when you take initiatives to get rid of error-prone, costly processes that don't add value to the end product in any way.

In terms of processes related to deploying their websites, Team Fun Dog is in full control over all changes to the TEST environment. That means that they can change software on webservers on TEST and that they have write access to the database server on TEST. They do not have write access to the database on LIVE, though. Access to the LIVE database is restricted to IT operations; in this example, Henry is their primary contact person. Team Fun Dog has a dependency here—they need IT operations, a.k.a. Henry, to be available to them when they deploy software to production.

In other words, when Henry isn't available, Team Fun Dog will not be able to deploy changes to LIVE servers.

Write Down Your Current Process

I'm taking the approach at this stage that you and your team have very little to no experience at all with mapping out your current processes for releasing software in your context. If your team has been working together for a while and have started quite a few releases together, the following exercise may not benefit you as much as it will a new team that has been assembled to take responsibility over maintaining and deploying legacy software solutions built by others. I'm starting simply to show how I would approach extracting information and knowledge if I were onboarding a new team where nobody had insights into how software would get deployed onto their systems. If, as you read, you don't see any value in doing this in your context, feel free to retrofit or improve the recipe outlined here to fit into your context.

The scene has been set. You have gotten buy-in from your sponsors and stakeholders for an upcoming automation initiative. You've been working with documentation for your current processes and have learned a great deal about cause and effect based on historical data regarding errors and incidents. Now the time has come. So what do you really do on a Tuesday morning at five after ten when you're gathered for a kick-off meeting and the leader of the meeting says, "All right, let's get to it then!"?

What you need to do is this:

1. Write down your current process for releasing software, from developer workstation up until production environment.
2. Visualize the process.
3. Add roles and responsibilities for each activity.
4. Add errors and incidents regarding build, deployment, and releases.
5. Make a sound plan of attack.

This is the high-level plan. You probably would like examples for "Write down your current process," which is perfectly fair. We should use Team Fun Dog as an example and draw a rough sketch of what they're doing today in a form that you should be able to quite easily reproduce for yourselves on a whiteboard.

Figure 8 is nothing more of a crude drawing of the infrastructure without much attention to details or anything, but it's a starting point nonetheless for many teams to create a drawing like this on a whiteboard. You have, in the example, a software developer in the top left corner who deploys changes from his developer workstation to TEST and LIVE environments. There are also storage solutions, such as a database engine for each environment, and dedicated search engine solutions per environment as well.

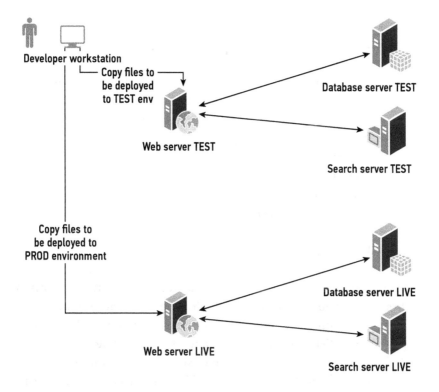

Figure 8. Rough, initial sketch of Team Fun Dog's environments and how digital deliveries are being deployed to LIVE.

In order to document the activities necessary for anyone to go through when deploying new features to LIVE, you should next write down everything you do in the form of a bulleted list, like Team Fun Dog has done below. Here, they describe their current, manual release process:

1. Log on to webserver TEST.
2. Copy files for deployment from my work station to a temporary folder on webserver TEST.
3. Copy files from temporary folder into website root folder.
4. If necessary:
   a. Adjust configuration settings in files.
   b. Log on to database server, make changes.
   c. Log on to search server, make changes.

5. Reboot webserver TEST to apply changes.
6. Validate changes.
7. Log on to webserver LIVE.
8. Copy files for deployment from temporary folder on webserver TEST to a corresponding temporary folder on webserver LIVE.
9. Copy files from temporary folder to website root folder on webserver LIVE.
10. If necessary:
    a. Adjust configuration settings in files.
    b. Log on to database server, make changes.
    c. Log on to search server, make changes.
11. Restart webserver LIVE to apply changes.
12. Validate changes.

Make a list like the one above for your team to get started. This is the first step of your journey. You don't have to do more than this to begin. It may even be completed within thirty minutes—that's good! Keep it simple. Write down what you and other stakeholders are doing in a bullet list and save it in the form of sticky notes, napkins, a whiteboard, or a spreadsheet. How output and knowledge is maintained isn't important right now; what matters is that you engage in a conversation and uncover hidden knowledge, which the writing process in itself will help you achieve. Describe work and activities that need to be done, and include as many details as you can.

> Pay attention to details. If it's absolutely essential for the success of a software release that you and your team open a specific file and validate it's content as part of your release process, it's vital that this action be present on the bullet list of actions that you're creating in this session.

If this exercise turns out to be too much of a theoretical exercise because you don't have the right people in the room, voice your concerns. It might be better to simply wait to do this step until the next time you have a scheduled release of software to your systems. While you're at it, have somebody sit next to the person performing the tasks to release the software. That second person can write down everything that's being done.

You could then tell the rest of your organization that you're engaging in a documentation initiative for your release process. Let them know that you might come and ask questions that you wouldn't usually ask to clarify nitty-gritty details about their current ways of working as part of a software release for the product in question. In low-trust environments, it's especially important to mitigate any elements of surprise by preparing your colleagues in other departments so the knee-jerk reaction won't be to become defensive or hostile against "you people" walking around with sketchpads mining for conflict and improper behavior.

This could be a great opportunity to reboot your current ways of collaborating, so use this documentation exercise to build relations with important stakeholders who you know will be affected by an upcoming change in release processes. Give first, ask later. Cake and a smile usually help to build good first-time impressions and show good intentions, especially if there is a history of noncommunication and bad karma between different departments in an organization.

Write down all your activities on a bullet list—this will be your first step.

Visualize the Process

Once you have built a common understanding of all activities involved, the next step is to create a list of all the parts of your infrastructure that you apply changes to. You will need to build a list of all your servers, all database instances, all index servers—in short, every piece of infrastructure that you directly invoke changes to as part of your release process should be on this list. To keep it short and simple, I'll use the term "node" to describe a generic piece of hardware or a software solution that you change or update as part of your software release process.

When looking at Team Fun Dog, they have the following list of nodes that are vital for their software release processes:

1. Developer workstation
2. Webserver TEST
3. Webserver LIVE
4. Database server TEST
5. Database server LIVE
6. Search server TEST
7. Search server LIVE

If you have nodes that share responsibilities, like a group of webservers in a load-balanced setup, it's perfectly fine to group these servers and have just one bullet to describe them. If Team Fun Dog had three webservers—A, B, and C—in their LIVE environment, it would be perfectly okay for them to just put one bullet for now and write "Webservers LIVE (A, B, C)" on the list of nodes. There's no need to be unnecessarily complex at this stage.

Once you have built your own list, you should then convert your list of actions and your list of nodes in your infrastructure to a swimlane diagram, where each swimlane represents a node or a group of servers in your infrastructure. See Figure 9.

If you and your team haven't spent time documenting your release processes before now, you will without any doubt gain knowledge about what you're really doing in the form of concrete, detailed actions when you're releasing software on your systems. There are likely to be activities that didn't make it into the bulleted list in the first iteration. That's perfectly fine and isn't a surprise by any means. Just fix any shortcomings as you go and improve the diagrams and the list of nodes as work progresses.

I have a single, silent prayer for you and your team at this stage: I strongly recommend that you do *not* digitize anything until you all have a high-level understanding of all activities that take place in your current process. If it's possible, plan sessions where everybody can be present in the same room at the same time. It's my experience that more people will engage and take ownership of the combined body of knowledge afterward. That's because the output, in terms of drawing the diagrams, won't reside with one person only—the single individual at the keyboard trying to tame Microsoft Visio.

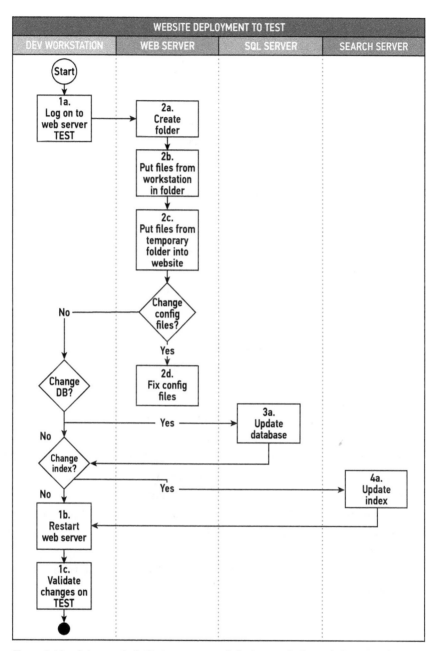

Figure 9. Visualising your bullet list is a great way of aligning everybody on which activities happen when. If you have lots of unknowns in your team about your current deployment process, drawing it will uncover a lot of hidden knowledge, which will come in handy in a subsequent migration process.

Even though everybody will be present physically and will be apparently equal in the process, that's not how it feels when one person is responsible for drawing the diagrams and list of nodes. Human beings tend to loose concentration and are unable to focus when IT friction inflicts interruptions in the flow of conversation.

I urge you to use sticky notes, pens, paper, and whiteboard markers at this point if you have any of these things available to you. It will greatly benefit the quality of the end result when everybody has access to the whiteboard and can add or remove sticky notes, erase lines, or expand a swimlane. It's also much easier to just move a string or draw a new line on a whiteboard instead of having to fundamentally change something in a large diagram already converted to ones and zeros when, after ten minutes of discussion, you find out that a process is different than you thought.

Wait on digitizing anything until everybody feels that every activity is present and you're discussing minor details unrelated to the big picture. It may take more than one session to get this far, but don't digitize until you reach this level of shared understanding between stakeholders. Once you've accomplished this, you should assign one or two team members to sit down and formalize your findings in a digital format so the rest of you can get other work done in the meantime.

If Context Is King, Unambiguity Is the Queen

As you can see in the swimlane diagrams, I've numbered each activity. The reason is that communication flows freely when there is no ambiguity involved. The flow of both synchronous and asynchronous communication, such as emails, chat messages, and the like, only improve when you add some means of identification to your activities. You can then write the following to your colleague:

> When we log on to the webserver in activity 12.C, I see the following problems emerge...

Compare that with the following, seemingly similar statement:

> When we log on to the webserver once deployment has finished, I see the following problems emerge...

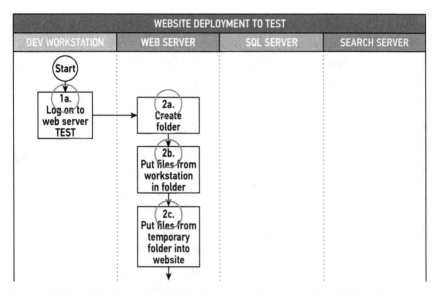

Figure 10. All activities in the developer workstation swimlane start with 1, all activities in the webserver swimlane start with 2, and so forth. Feel free to devise your own scheme if this approach doesn't provide a good fit for you in your context.

There is a built-in risk for the receiver to look at the wrong activity in the diagram, especially if the diagram gets complex with lots of activities and nodes involved.

ID tags for your activities can be designed in multiple ways. You can have a number that just rises in increments, but you'll get tired of doing so very quickly when diagrams evolve. I usually go with a combination of letters and numbers, assigning a letter to each node or swimlane and then a number for each activity targeting that node. This makes it easy to expand the diagram when you add new activities new activities without changing the ID for all subsequent activities across all nodes (see Figure 10).

All activities have an ID, but if you look carefully, the second item in your bulleted list from the very beginning has been refactored into two distinct activities here—one focusing on creating a folder, and one where the files are being copied into the folder. It's easy to identify individual, independent steps like this since there might potentially be some knowledge or domain logic embedded in a seemingly simple activity.

Let's take the example of creating a folder in activity 2A in Figure 10. Are you allowed to name the folder whatever you like, or are you expected to adhere to a naming convention? Where should the folder be located on the server, and should everybody have access to it? The devil is in the details. If your team involves more than two or three people working together describing and visualizing your release process, it's best to slice up one activity into as many steps as you need to express and document the necessary knowledge and metadata. There is no secret formula describing how deep you should go while doing this, but ensuring that all activities—no matter how many you need to create—have a unique identification tag increases clarity and the quality of communication afterward.

It is vital that all activities are well described and all individual nodes, or groups of similar nodes, are represented in a swimlane. You may dive deep into activities that you and your team are fully responsible for end-to-end, but the granularity of an activity is less important for activities where you are not in full control of execution. At this point it's perfectly fine to just have one single activity in swimlane describing work being done by others even if you know that there are multiple, independent steps executed on your behalf.

Let's say that Team Fun Dog knows for sure that Henry the database administrator executes multiple scripts as part of updating the LIVE database. They could describe each script execution as a single activity, but I would recommend having just one single activity describing the fact that IT operations executes database changes, as long as the chain of activities will always be the same. You may combine and aggregate activities into larger abstractions as you wish, but all decision paths should be present at all times in your diagram.

Let's say for instance that Henry executes a series of scripts on behalf of Team Fun Dog, but on different databases. That would perhaps be very important information to include, especially if database A and B should always be updated as part of a release, but database C isn't necessarily to be touched at all *except* on the rare occasion when they have updated a well-defined set of specific database tables. When Team Fun Dog does, however, it's absolutely crucial that Henry validates the conditions of the recent backup, since Team Fun Dog will be unable to perform a rollback in the case of a failing deployment. They tried that six months ago, and it cost them two hours of downtime plus having to manually clean up duplicate data and buy everybody in IT support cake for a week.

When you start writing your diagrams, you don't have to maintain the bullet list you made in the very beginning. The bullet list was just an easy way to get going and achieve something quickly, so you can throw it away once you've started working on your swimlane diagrams.

What's important going forward is to continuously maintain and update the diagram once you've made your first end-to-end draft in digital format.

This is an example of hidden knowledge that you may want to express in your diagrams. It isn't enough to have one single activity saying "Henry does stuff" because, depending on the context of your upcoming release, Henry may need to execute a different set of activities. This seemingly hidden knowledge should be expressed in your overall diagram, so all decision paths are covered fully—including activities that you aren't fully responsible for executing. You shouldn't go as deep as to document and understand unimportant details of execution and how others do their work, but explore the nature of the activities enough to understand what's going on at a basic level. Keep an eye out for words and phrases like "unless" and "it depends" when you discuss what work is being done. In the sentences that follow words like these, you will find invaluable information and assumptions about your specific context that you need to be explicitly aware of and document in your swimlane diagrams.

Using a Parking Lot to Maintain Focus

You and your team have by now gained a deeper understanding of activities being performed as part of your software release processes. Maybe you've just explicitly confirmed what you already knew. Maybe you have gotten a few good ideas about how you can improve what you're doing today based on earlier experiences and conversations.

Those are likely all good suggestions and ideas for improvement, no doubt. However, I will recommend that you don't jump straight into solving these issues before you make a sound plan. One reason is that perhaps the problem you experienced won't ever happen again. The process will likely change significantly in a short while due to the introduction of automation or similar initiatives.

Figure 11. Example of a parking lot.

There are other good reasons for focusing on the task at hand, so if you get good suggestions during your conversations and sketching sessions, write them down in a parking lot—whether it's analog or digital isn't important. The insights are very valuable and will come in handy at a later point, but maintain focus for now. You will get back to your great idea for improvement later on; just write it down and put it on hold for now. Focus for the time being on ensuring flow in the documentation phase you're in. The parking lot is a simple tool for keeping discussions on track and focusing on the job at hand.

Add Roles and Responsibilities for Single Activities

You and your colleagues will likely create more diagrams, perhaps one per environment per product, and perhaps a combined diagram where all environments are present for all environments. You have an overview now of what's happening, from a developer's point of view, up to the point where a change has been deployed to your LIVE environment. Well done!

What you should spend some time investigating now is who's responsible for doing what. If you and your team knows everybody involved, this exercise won't take a long time, and you might not uncover new knowledge.

But there's a very profound point in explicitly declaring on a diagram where responsibilities for execution are transferred out of your team or department. When you transfer responsibility by handing over tasks to other people, you are forced to wait for others to prioritize your task and execute it. Unless you can do other work in parallel in context, you either wait for completion or you switch context and start working on other work in other contexts. Regardless of what you do, your team will experience waste in the form of waiting or task switching. When you have activities that are likely to incur waste in terms of misalignment, lack of information or clarification correspondence back and forth, it is vital to have those activities documented for tracking purposes. You and your team would be able to measure waste explicitly if you could count the number of hours spent actively working on executing activities related to software releases and then look at the number of hours spent waiting for others to complete the tasks you've handed over to them. The waiting time shows the overall efficiency of your deployment process, which is why you want to know exactly where and when tasks are handed over to others to execute on your behalf.[8]

Let's go back to our database administrator Henry, who is the only one with access to the databases that Team Fun Dog are using on LIVE. This organizational setup means that Team Fun Dog has to create a ticket or change request to get changes deployed to their databases, which is a potential roadblock and an event where the human factor should never be underestimated. If Team Fun Dog communicates incomplete or flat-out wrong information in the change request they create for Henry in IT operations, he may either waste precious time writing back for clarification or he may proceed with incomplete information at the risk of failing his part of the job. That could abort the release, postpone it, or something in between, while time just passes by. Stress and tension will ripple through teams and departments involved in handling the unfortunate situation unfolding in front of them.

You should add knowledge about responsibility transfers into your swimlane diagrams to visualize risk in the form of human factors, undocumented processes, hidden knowledge, and lack of adequate prioritization of tasks on the receiving end of your activities. Every risk is an opportunity for optimization in the form of automation. In the case studies at the end, I'll provide a few proposals for a solution that will take different needs into account and have different implementations for solving or mitigating the same problem.

---

8 "The Seven Wastes: Seven Mudas," *Lean Manufacturing Tools*, accessed July 15, 2020. https://leanmanufacturingtools.org/77/the-seven-wastes-7-mudas/

Look at your swimlane diagram(s) one more time. This time you should extend the diagram with a bit of coloring, like this:

1. Give activities where you and your team are in 100% control of execution a green color.
2. Give all activities where you aren't fully in control a red color.

If you are having doubts whether an activity is green or red, this is a signal for you to investigate why this is the case. Then you can slice a potentially mixed activity into distinct green and red activities.

Team Fun Dog has a dependency in their release process toward Henry in IT operations, as depicted in Figure 12. You may have many more dependencies, or you may have none at all because you are in full control from A to Z of your software release process. In either case, it is vital that all dependencies are known and can be communicated to external stakeholders afterwards.

....

Entry and Exit Criteria

Your diagram is slowly evolving. You and your team started out with a list of vaguely described activities that you have reworked and enhanced into diagrams with distinct activities and metadata describing context such as dependencies toward external stakeholders. The remaining pieces of information still missing are entry and exit criteria,[9] which you may also know as "pre" and "post" conditions. Naming isn't important, but for the sake of consistency, I'll use the terms "entry" and "exit" criteria going forward.

Entry criteria are the conditions and data that an activity requires in order to execute and report back to the workflow that an activity has completed. In the case of Henry the database administrator, he needs to know:

---

9 "Entry and Exit Criteria," *What Is Six Sigma*, accessed July 15, 2020. https://www.whatissixsigma.net/entry-and-exit-criteria/

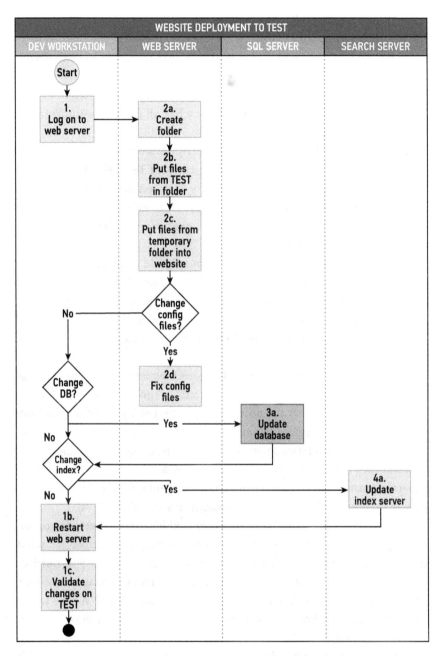

WEBSITE DEPLOYMENT TO TEST

| DEV WORKSTATION | WEB SERVER | SQL SERVER | SEARCH SERVER |

Start

1. Log on to web server

2a. Create folder

2b. Put files from TEST in folder

2c. Put files from temporary folder into website

Change config files?

No

Yes

Change DB?

2d. Fix config files

3a. Update database

Yes

No

Change index?

Yes

4a. Update index server

No

1b. Restart web server

1c. Validate changes on TEST

Figure 12. Team Fun Dog has a dependency towards IT Operations doing database updates on their behalf. Visualizing dependencies is best done using contrasting colors such as green and red.

- Which database should I update?
- Which scripts should be executed on the server?
- How would a successful response look once the script has executed?

These are only examples. There may be more relevant information needed before an external stakeholder like Henry has all necessary information available to initiate a set of activities.

Exit criteria, on the other hand, are descriptions of data that the activity should return before it can execute successfully. If Team Fun Dog needs to have certain information back from Henry to determine what they should do next, this information should be regarded as exit criteria that Henry should provide as part of his successful execution of the activities he's responsible for in the flow of events.

Another exit criteria may be that output should be located in a certain place, such as a note on a support ticket that Team Fun Dog has created for Henry in order for him to document the work he needs to undertake on their behalf. An email with the output attached may be sufficient, but if that isn't good enough, what should the format of acceptable output from an activity be?

If Team Fun Dog in its context is the correct owner of the release process yet don't know these conditions at this point in time, they are responsible for uncovering unknowns and documenting expected input and output from activities. The same applies to your team in your context. The accountability for aligning expectations rests with the team responsible for the release process. They are the ones who should take initiative to document and align expectations throughout the entire chain of events, from developer workstation up until a release has been validated successfully on LIVE.

Exit criteria from an activity may very well be entry criteria for other, subsequent activities. In your situation you may ask yourselves what information you need to hand over, or what information you need to ensure is available to forthcoming activities for them to execute. Likewise, what do you expect from an activity that it should return? Where to? In which format? To whom? On which channel—email, Slack, ticket system, or something else?

It is crucial to understand that even though you may not be responsible for all activities because you have external dependencies, you are in complete control over entry and exit criteria for each activity that you have described. You *must* take responsibility for explaining and motivating stakeholders as to why, how, when, and in which format you need information from them. You may adjust or refine your initial suggested entry and exit criteria as you talk with stakeholders and discuss do's and don'ts along the way. Let them have their say as well taking responsibility for the overall result—as long as you always remember that the end-to-end accountability for the correctness of your swimlane diagrams lies with you and your team.

If it makes sense, you may conduct entry and exit criteria for all activities, green as well as red ones. It may potentially provide you with valuable insights when you systematically work your way through all preconditions of the work being conducted inside each activity. There is a chance that you decide it won't bring you any value or reveal hidden knowledge. This is a valid conclusion, and you should adapt and spend your time wisely, of course. In any case, you ought to establish entry and exit criteria on your red activities to explicitly align expectations. No matter what happens, you will need to communicate without ambiguity and with an acceptable level of trust when you request that others take on extra work or do work differently in order to comply with expected exit criteria once their work is done.

Summary

When you can show how you are able to express what you need from other people and roles and why you need specific bits and pieces of information from them for your team to succeed, you will likely see an increase in trust between teams and departments involved. Communication between organizational silos can flourish in ways that seem unimaginable in some organizations if you approach your automation initiative with respect for other people and take their valid viewpoints into account.

Success isn't guaranteed just because you are able to express and document your current processes before an implementation effort begins. You should also prepare yourselves for the fact that it takes time and requires a fair amount of focus to generate and uncover treasures of knowledge. Celebrate your small successes along the way; short-term wins provide motivation for further investments and build foundations of shared knowledge that you will leverage upon going forward.

Call in for a thirty-minute demo of your diagrams if you have a session with particular stakeholders who you depend upon. A guy like Henry the database administrator will benefit from understanding the entire chain of events leading up to the ticket, which will be his first and last touchpoint in the entire process. Stakeholders may not express this themselves or show any signs of enthusiasm whatsoever, but they will recognize the value-add of having seen the entire process leading up to the work that you impose on them from time to time. Maybe not to your face, but you and your team will get increased respect and gain trust in your organization for being able to show that you know what you're doing and where you want to go. This type of leverage will come in handy when you begin to change and automate processes in a short while.

The diagrams you're finalizing now will need to be revisited from time to time. They're not static in nature, since they reflect your ways of working, which will change over time too. Going forward, you may need to revisit the work you've done so far, which is perfectly fine. New revelations will emerge, new technologies and trends will be picked up, and they will all in turn influence how you do business. The diagrams should follow, and it is up to you to ensure that this happens regularly.

Rest assured: You're walking in the right direction. Keep going! Give your team a high-five and allow yourselves a break from time to time. Join forces at the coffee machine to loot liquid supplies for the long hours to come.

Part 3

How to Build Your Software

DURING THE LAST FEW CHAPTERS, you've built toward the work that you and your team is about to plan and execute.

I've found that across technologies and operating systems, there is a pattern of recurring activities with technology-agnostic entry and exit criteria taking place in just about any automated workflow that is building and deploying software to nodes in an environment. You and your team should carefully consider how you wish to implement each activity in your context. Part 3 doesn't describe a concrete implementation strategy for this series of activities but rather an architectural approach and a set of considerations you and your team will need to take into account before deciding how to implement an activity.

My goal for doing this is to separate the solution from the constraints of technologies, since scalable and maintainable architecture suggests that you continuously allow yourselves to evaluate and eventually replace technologies as your software architecture evolves. This is how enterprise architecture and software architecture should be viewed: as a set of roles and responsibilities, regardless of how those roles and responsibilities are implemented in terms of the technology used. This is the framework for the following chapters.

> It is beyond the scope of this book to describe technology-agnostic design (TAD) and technology-agnostic architecture (TAA) in depth. I recommend that you and your team look at the list of books and articles in the appendices if the concept of TAA and TAD looks promising in your context.

The Single Most Important Thing in Beginning an Automation Initiative

A lot of things depend upon the success—and potential failure—of your upcoming automation initiative. Organizational and team culture is probably something you will need to work with as part of implementing automation going forward. For example, you will likely need to increase competencies in test automation because it will introduce friction in your team when people need to change their behaviors and work differently.

There are also prerequisites in terms of technology and ways of working that are important. Sharing hardware is a fallacy which I will describe in detail later on, but that's not the biggest challenge if you engage in automation.

When I read books on Continuous Integration and Delivery describing prerequisites and best practices, I feel that there is a shared understanding across the industry that source code is always placed in a version control system. It's like, "Water is wet, the sun is hot, and source code only lives in source control systems." When I go and see what some teams are doing, though, the picture blurs a bit. I have come across teams that develop solutions for years without having code in version control. I also find that IT operations departments especially, with highly skilled, highly competent engineers, simply have never been introduced to version control and the benefits of having their custom script files stored in a version control system rather than a fileshare. It puzzles me, to be honest. Once you've coded yourself into a corner and have tried rolling back to where you began two hours ago to start over without having to figure out for yourself what you in fact did change in various places in a 2,000-line file of legacy script, you're hooked for life, I can tell you.

I couldn't imagine writing any code without some means of source control being available, even if it's just hacking together a script for my own use. I can tell you this: If there is one single prerequisite determining the success or failure of any automation initiative across technology stacks and team competencies in automation, it is whether or not you have your source code in a version control system. If you don't, you will fail. No, let me correct that: You won't even be able to begin if your source code cannot be retrieved from a version control system.

You cannot automate your software deliveries if you cannot instruct your build servers to download a given version of the source code from somewhere; it's as simple as that. This may seem quite obvious to some, but it isn't common knowledge, in my experience, if teams and organizations have zero knowledge about Continuous Integration and Delivery automation. Maybe those of us who have been using version control for decades are cursed by knowledge and simply don't realize that there was a time where the concept and importance of version control was an unknown-unknown to us. What teams and organizations learn by looking at Continuous Integration and Delivery best practices on the internet is that, say, trunk-based development is the right pattern to use (which it is) or that you should commit code early and often (you should indeed). But the underlying assumption here is that the source code is *already* in a source control system of sorts when teams engage in automation initiatives. In my experience, that's just not the case everywhere. That's why I'm taking the time to state explicitly that you will go nowhere fast if you try to automate deliveries of source code that doesn't reside in a version control system.

Going forward, I'll assume that your code is located in a version control system of some sort; it's not very important for now which one. It doesn't mean that you should have all your digital artifacts, such as images, database schemes, and stuff like, that in version control before starting out; it means that the flat text files with source code used to build and subsequently deploy state changes to nodes in your environments originates from a version control system containing a history of changes to each file in question.

Which Platform Should We Use for Deploying Software?

All automation projects begin with a version control system and the toolset that you have available for the upcoming implementation. That means we will need to spend some time looking at the Continuous Integration and Delivery platform that you and your team will be working on going forward.

Often teams end up having lively discussions about which platform is the right one. The definitions of *right* and *wrong* depend heavily on what questions you are asking, so I will emphasize the importance of asking the right questions from the very beginning. If you are able to choose your platform freely, it is a great advantage that you should feel grateful for because it allows you to replace your existing platform if you deem that change necessary to achieve your goals. The opposite situation—where you're confined by a business strategy demanding that everybody use the same enterprise tool that others before you have evaluated and bought but doesn't provide a good fit for your needs—may potentially be a roadblock for your automation initiative.

I've personally been part of a team where two highly skilled professionals tried for several weeks to automate building an iOS app on an enterprise system that by no means was built to handle the iOS operating system. It never really worked in a way that customers (i.e., app development teams) found useful, and it was end to end a poor and expensive project that didn't provide value—except as a learning experience for the rest of us. We learned how dangerous it is for an organization to force different teams with different constraints into a one-size-fits-all system not built with extensibility in mind.

If you and your team find yourselves in the fortunate situation of having both proper permission and the responsibility of choosing a platform suitable for your needs, it is a great advantage for you going forward. I don't want to go into too much in detail here, but you should go online and check up on the market for solutions that provide the best fit for solving your current pain points. I won't recommend a list of names and technologies in this book because I'm writing these words in May 2020, and if you read this in three, four, or five years from now, technologies and trends will have shifted tremendously. Whenever I mention specific companies in this book, it will be in the form of examples providing context for specific arguments.

I can guide you, though, toward foundational principles and considerations that you should be doing regardless of scenery. I'm betting the rules and principles uncovering risk and assessing a market across technology stacks will be just about the same five years from now as they are today. There will always be a framework and a series of questions you will need to ask yourself as a team and as an organization when you buy new software. In the following chapter, I offer some basic questions as a way of getting started or perhaps as an add-on to your existing vendor selection process in your organization. You can look through the questions that I consider important and make up your own mind if they would make any sense to investigate in your own context.

Which Integrations Do We Need?

You want flexibility and options once you have decided to invest in a platform. Your solution should be able to integrate with technologies that you're already using today. If your organization, for instance, uses a series of enterprise products from a single enterprise company, it is evidently an argument for the advantage of candidates who have built-in seamless integrations with those systems. If your employer has a strategy for using wall-to-wall Microsoft technologies on strategic initiatives, it kind of defeats the purpose if you buy a nonstrategic, third-party solution that doesn't integrate very well with the Microsoft platform. You may not have a use case today for integrations and data exchange, but believe me—that need will arise somewhere, somehow, and sooner than you might think. It is at this point, when systems interact with each other, that you and your team will benefit and shine in the eyes of your organization if you are able to swiftly honor demand for integrations that weren't part of the evaluation scheme when you were assessing vendors and qualifying candidates.

The way you achieve this level of agility is to relentlessly pursue buying systems that are built with the intent of extensibility and integrations with other systems. The best examples are built from the bottom up using a so-called plug-in architecture where the entire platform is built to retrofit bricks of different functionality, each able to expose data and extend upon standard functionality. By allowing others to provide necessary third-party integrations to other systems, you let other companies do the work of building and maintaining solutions that solve a particular need for a fraction of the customers on your platform in the form of extended reporting features. It doesn't matter much in most cases if those add-on integrations provided by third-party vendors are opensource or licensed versions. Look for quality in terms of things like performance being acceptable, the technical documentation being updated and providing decent self-service capabilities, and support being available, such as an active community responding to bug requests or documentation clarification issues.

If you and you team decide in favor of a specific vendor and end up churning out custom code to achieve basic needs for integrations, you've chosen the wrong system. Don't go any further down that road. You will end up spending a lot of money on a broken system that will be impossible to upgrade and where the speed of onboarding new employees will grow proportionally with the number of lines of custom integration code you put into your platform. Across all parameters, it will be a bad choice to proceed, especially if the problem has already been solved by others. You will very quickly break even if you spend $1,000 for a license if you calculate total cost of ownership on your custom-coded solution, believe me.

A classic example of a must-have integration is the ability to configure single sign-on. Given the prerequisite that you have federated login systems such as Active Directory Federation Services (ADFS) or similar already, it should in my opinion be a roadblock if a candidate for a Continuous Integration and Deployment platform in your context isn't able to authenticate users against your federated login system.

Another example of a must-have integration would be out-of-the-box integration with your source control system. I say that maybe you're using an outdated version control system if the platform cannot integrate without doing customizations but in all other parameters seem to be a mature product. Big cloud companies such as Amazon AWS or Microsoft Azure have built-in version control systems in their portfolio of products supporting software release management. If you go for platforms like these and have your source code located in a legacy solution elsewhere, it might be worth considering the possibility of migrating your source code into the platform of your choice and aligning your portfolio of systems that you use in day-to-day operations. Why pay licenses for third-party products handling source code versioning if version control could be handled by storing your source code in the Continuous Integration and Deployment platform of your choice? Those are questions that you and your team should be asking yourselves. If nothing else, you should be able to communicate and ask why the organization should pay for licenses and operational costs of having multiple systems when only one of them seems necessary.

These are only examples of integrations to get you started. There are many, many more available. You, your team, and your organization will probably have lots of other ideas. Remember though that nothing is free, so focus on the integrations you absolutely cannot live without, and don't implement any that don't provide business value in your context. If you're uncertain about which integrations may become necessities, put ideas for integrations into the parking lot mentioned earlier and revisit the list of parked items from time to time.

How Much Does It Cost?

Asking about cost is the wrong question. You don't want to know at this point what a platform will cost you. Let me rephrase a bit: You don't *need* to know at this time what a platform will cost you. What you're currently doing is mitigating and assessing risk prior to buying a platform that will be of strategic importance to your organization and your ability to compete in the market.

If your ability to deliver high-quality software is lessened by your buying into a certain platform, will you be handing over a competitive advantage to your direct competitors? I would say yes. Is the price of any given platform the most important thing if your platform could provide *your* organization with competitive advantages? You would potentially increase the agility of your organization, allowing teams like yours to respond to changing business requirements without sacrificing quality because the platform was built to support the notion of constant change. This enables you to change direction and enable technology adaptation when advantages emerge. I would say no, the price tag isn't the most important thing in this case.

Price matters of course, but the price of a Continuous Integration and Delivery platform shouldn't be considered an expense. It's an investment more than anything else. At this point in the process, you and your team are likely to be in a phase of clarification and scoping of requirements, which means that the price tag is just one out of many factors. You should take a more risk-based approach and ask yourselves, "What's the likelihood of {insert risk} happening?" and subsequently "Okay, how much will we then invest in mitigating the risk of {insert risk} happening?" You have a much better chance of answering those two questions rather than trying to estimate a total cost of ownership at this point in time. You can only guess at this point what the cost of onboarding and education will be in the long run anyway, even if you get fixed price quotes for a team of consultants who will fly in, train, and certify you and your team members before they drive back to the airport two days later. Learning how to do Continuous Integration and Deployment is just like learning how to drive: It's not the theory that's the hardest part. It's the first time you're driving alone in your car after a snowy night and you feel the car responding differently when you apply the brakes that you start to really learn anything.

If you need a framework for discussing price, I would suggest that you use an approach such as Decision Analysis and Resolution (DAR). In short, it's a technique used to analyze possible decisions using a formal evaluation process. You establish a set of criteria used to evaluate identified alternatives and then apply a score to each of these alternatives using a matrix where all criteria are listed on the Y-axis and each proposed solution or decision is listed on the X-axis. The scores you apply are whole numbers in a range that you decide. The range may be from 1 to 3 (poor/average/good), 1 to 5, or whatever you decide makes sense in your case. As a rule of thumb, go with a scale of 1 to 5 to avoid non-value-adding discussions about whether or not a vendor score should be a 6 or a 7. Nobody knows for sure whether or not a score should be a 6 or a 7 anyway. (Doing a Five Whys on "We can't decide if Vendor A is a 6 or a 7" will probably reveal that somebody on the team has a really hard time finding their peace with uncertainty at work, but that's different story.) I would suggest that you use a range of no more than five numbers.

An example could be buying a family car. You probably have constraints in the form of a budget and that the car should be used for specific purposes, such as driving kids with muddy feet and stuff like that. You can use a Decision Analysis and Resolution matrix to drive the conversation forward and compare alternatives.

You can see an example of a DAR matrix evaluating Continuous Integration and Delivery platforms from Vendor A, B, and C respectively.

There is a bit of work to do regarding the decision criteria. Seek inspiration on the internet; there are loads of ideas out there. Invest some time here, but restrict your effort so you don't evaluate fifty or even hundreds of criteria. The matrix should help you choose, not become a meta-project in itself—if it does, you're doing it wrong. Stick to five to ten criteria for comparison and a maximum of four vendors.

Criteria	Vendor A	Vendor B	Vendor C
Version control support	1	2	1
Docker support	1	3	2
User interface look & feel	1	1	3
Integration with public clouds	3	3	1
Active community	3	1	2
Extensibility via 3rd party plugins	2	2	1
Initial cost	3	2	1
Sum	14	13	11

Figure 13. Example of a Decision Analysis and Resolution (DAR) Matrix.

You will need to agree upon the weighted scores you give each vendor. If you deem it important to only use tools with an active community, how do you translate "active community" to a number range between 1 and 3? What is the difference between a 2 and a 3 in terms of "initial cost" in Figure 13? You need to have that conversation up front before applying numbers.

Cost isn't of much interest to you at this point; it's just one factor out of many. It's more complex than that one parameter to assess whether or not a specific vendor will be the right fit for you. Using an approach such as DAR to assist you in your selection process gives you a framework for discussing facts and numbers instead of feelings, which will benefit the quality of the decision process by lengths.

Where Are Data Located?
The European Union (EU) implemented back in 2016 regulation 2016/679, also known as General Data Protection Regulation (GDPR). It concludes data such as your name, phone number, email address, Social Security number, etc., that you hand over to a company as part of a business relationship but that belongs to you and not the company in question. It is the responsibility of companies to ensure that these personally

identifiable data are stored in a responsible fashion, i.e., by anonymizing data. Companies also commit themselves to being able to delete data upon request and classifying data so that it will be deleted on a retention schedule when they are no longer needed by the company in its line of business.

Even though the regulation has been approved by EU and implemented by each of the union's members, the consequences of GDPR are in effect global because all companies and organizations doing business within the EU obligate themselves to adhere to the GDPR regulation. It means in effect that if you, as an American or Chinese company, sell a product or a service to a customer living in the EU, then this company agrees to the code of conduct laid before them in the GDPR regulations. The company should be able to give the customer a complete list of its knowledge about the person when requested to do so and upon request completely discard all data that belongs to the person from their systems.

The member states of the EU have focused quite extensively on privacy legislation and the fact that ownership of data belongs to the individual handing over data and not the company in charge of storing it. The EU's court system uses the stick on companies not incorporating the legislation as it is implemented in each member country. The legislation in itself does not describe in detail how GDPR should be implemented in each country, so there is room for interpretation in terms of severity and what constitutes a reasonable fine.

Examples, though, tell a story of how harshly countries in the EU have interpreted the GDPR framework. A case testing the boundaries of GDPR from Denmark is a good example. A taxi company called 3x34 received a fine of approximately US$200,000 for not deleting personally identifiable data regarding 8.8 million taxi fares.[10] The company in question defended themselves based on the fact that the name of the customer had indeed been anonymized, but during an audit the auditors revealed that phone numbers did not undergo a similar anonymization exercise. Hence a taxi fare could be connected to an identifiable person long after the taxi fare had been completed. At the time of this writing in May 2020, 3x34 has

---

10 "Recommended Fine of DKK 1.2 Million to Taxa 4x35 for Violation of the GDPR," *Gorrissen Federspiel*, March 27, 2019. https://gorrissenfederspiel.com/viden/nyheder/recommended-fine-of-dkk-12-million-to-taxa-4x35-for-violation-of-the-gdpr

not yet gone to trial in the court system, but the case indicates how GDPR regulation is being interpreted in membership countries and how much scrutiny you're under as a company in terms of storing and handling confidential customer data.

Other countries have felt the harshness of EU legislation when they have not been able to comply with GDPR. One of the details of GDPR is the fact that the EU has strict opinions about where data is stored physically when you are doing business within the EU. In short, data concerning European citizens must be stored in a datacenter physically located in one of the EU's membership countries. This circumstance did indeed cool down the relationship between the EU's political leaders and American tech giants such as Facebook, Google, and Amazon for an extended period of time. Their infrastructure and systems were never built to take into account that idea that data in one datacenter should only ever be mirrored to other datacenters around the globe if the person owning the data was in fact a citizen of the European Union. The response from EU after a long stretch of negotiations has resulted in nine-figure fines in US dollars, which clearly confirms that the European Union has taken it upon themselves to use their political powers to push for a different approach toward privacy than what has been prevalent since the birth of internet.

Some companies use GDPR as an argument in favor of on-premise solutions. If all data resides in a datacenter located in office buildings you own, you are certain where data is located. That is correct, but you're not done just because you focused on the physical location of storage facilities. If you don't as a company gate access to data and you have no policies for data retention, you are likely to end up in trouble at some point. But that's a different story altogether.

The premise of this chapter isn't to carve out every little detail about GDPR. It's a matter of framing where you should focus your efforts when you engage in the vendor selection of a Continuous Integration and Delivery platform. The physical location of databases may be a roadblock for onboarding one vendor's solution in favor of another due to the complications around legal compliance and customer expectations as regards your abilities to handle their personal data in a responsible fashion.

The story of 3x34 and others serves as an example that you should take privacy and data protection seriously. You and your team must not underestimate the potential consequences of a Software as a Service (SaaS) solution you may have the mandate to purchase but where there may be a series of legal complexities in terms of ensuring that your company remains GDPR compliant.

If you and your team are part of a larger enterprise operating within the EU, it is likely that you have already experience with GDPR regulation and how it should be implemented. Ask around if you have knowledge within the company already; otherwise, spend some time investigating if GDPR is a factor that you need to account for. As you can see, it may have severe financial consequences not to be in complete control over data persistence and data retention. A typical problem in your case may be an implementation of single sign-on, where user profiles with information about names, email addresses, phone numbers, etc., are mirrored to each system using your federated login for authentication. If each of those systems are not GDPR compliant for one reason or another, it may pose a risk that you should at least be aware of and prepare to mitigate.

Making the Final Decision to Select a Platform

The picture may seem a bit bleak because the EU is capable of fining companies for billions of US dollars, but I would still advise you to take calculated risks along the way. Avoid analysis paralysis in trying to think about the risks of every possible outcome of a decision regarding the choice of platform. It will only lead to frustration. It's a much better approach to try something, which enables a feedback loop of empirical evidence.

Of course, you shouldn't go with systems that store passwords in clear text or build your own solution. That's not the case here. If you find two different vendors that both look suitable and both get decent scores in your DAR matrix, try them out and see what they are worth. It's better to inspect and adapt after trial period of thirty days. During that trial, focus on learning and maybe trying out how easy or difficult it is to navigate the system. See how much time it takes to integrate a third-party component before you commit.

Terminology for Building a Software Delivery

Before we move on, I would like to walk you through a few terms that are also found in the glossary and here dive in a bit deeper. You will encounter these terms quite often for the remaining part of this book. It's an epic mistake in communication to not settle upon shared interpretations of the meaning of certain domain-specific words, which is why I'd like to be explicit about terminology before we proceed.

> I recommend that you look up words and terminology in the glossary in the Appendix.

Artifact

An artifact is a folder or a zip file containing files needed during a deployment of software to an environment. It contains the bill of materials for your upcoming deployment to nodes in your infrastructure.

The content of an artifact may be binary files, HTML files, configuration files, script files, or graphical elements such as images, sprites, and the like. It makes a lot of sense to also include other fragments of knowledge and metadata to provide context for your digital delivery. You may have documentation or release notes for your digital delivery, and you should include these explicitly in the artifact's representation of the actual delivery. From a process point of view, an artifact is the single source of truth where people and systems should always be able to find all knowledge concerning a digital delivery. You don't want bits and pieces of knowledge spread across wiki pages, folders on network

Figure 14. An artifact is a bill of materials for your software release.

shares, SharePoint lists etc. It is of course likely that an artifact contains references to data, such as a list of URLs in an XML or JSON file, to avoid the time and size penalty of copying blocks of data into your artifact all the time. The important notion is to consider an artifact as your single source of truth concerning a specific software delivery.

The example I tend to use when explaining what to put into an artifact of meaningful knowledge, besides the actual delivery itself, is a small file I always create called "version.txt." This file is generated as part of building the software, and I usually place it in the root of the artifact folder. It contains the version number of the build that the artifact represents and nothing else. This file comes in handy when you're logged into a node that doesn't behave as expected after a deployment. You need to quickly determine whether or not you've deployed the correct artifact or if the cause of the error you're investigating is to be found somewhere else. It's also very easy to have machinery and computers scan the contents of version.txt and validate what's in there against an expected value. It could raise an alert if the deployed version does not match the one that the monitoring system expects to find.

A deployment may involve more than one artifact, where each artifact represents one part of a combined set of materials you want to deploy into production. Best practice is to have an artifact contain all the files you will ever need to deploy on all environments. You may choose to have one artifact containing your website and one artifact containing the configuration files alone, with the extra overhead of having to support and maintain dependencies between artifacts at the time of deployment to nodes. It is up to you whether or not you want to take it upon yourself to handle extra layers of abstractions in terms of dependency management. If you're new to Continuous Integration and Deployment, I would suggest having as few dependencies as possible. My advice in this situation would be to include everything, including your configuration files for all environments, in the same artifact always.

I'm well aware that this contradicts best practices in terms of designing build and deployment pipelines. The argument in favor of splitting deployment of the artifact and the configuration of a deployment is that in reality the two bills of material are indeed two different deliveries that ideally should be able to be deployed independently. When you put everything into the same artifact, you will need to not only reconfigure your solution when you wish to reconfigure your solution, you will also have to actually build your solution from scratch, producing a whole new bill of materials. This is probably what you really want to do in the first place. For that reason it is considered antipattern, and I do agree. However, I would still recommend that you take the pragmatic approach of having everything in one artifact if you have little experience in automating software deliveries. The penalty of managing dependencies between artifacts is, in my opinion, not worth paying at this point on a learning curve.

You and your team will, at a later stage when you have built up a foundation of knowledge and experience of delivering software using Continuous Integration and Delivery principles, be able to chop up a large, monolithic artifacts into smaller, independent lumps that potentially could be built and deployed through dedicated, independent build and deployment pipelines. It would be natural at this point to create a dedicated artifact only containing the bill of materials needed to configure a deployment of a service or a set of services without having to perform an update of the actual service in itself.

I usually recommend that you build the structure of your artifact in a way that allows for configuration and other deliverables to live in self-contained folders. In the example below, a website has configurations packaged in separate root folders:

- Website (your website files, including global configurations that don't change regardless of target environment)
- Configurations
    - Test (configuration files for TEST environment)
    - Live (configuration files for LIVE environment)

This will make it much easier at a later stage to split up this large artifact into smaller artifacts if you need to. Another benefit of adhering to this pattern is that you will gain very detailed insights into how your website or delivery is configured when you explicitly need to grasp configuration files in a solution that will require changes for the entire delivery to work properly.

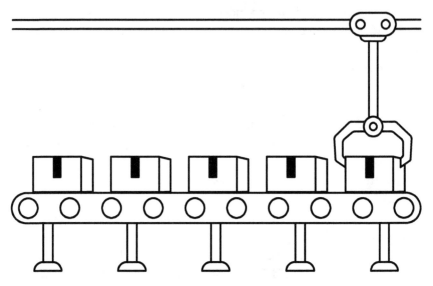

Figure 15. A build is merely an assembly line of activities producing an artifact.

Build

A build is the set of activities executed on a virtual assembly line producing the bill of materials that will eventually be part of the output in the form of an artifact.

Imagine a warehouse where you to begin with an empty pallet and a pallet truck. You have a package list, so you walk around plucking products down from your shelves putting it into the large cardboard box on your pallet. Once you're done, you seal it with duct tape and apply a unique number for identification purposes. The pallet with the box is set aside in a known location. You will retrieve it again tomorrow when the box should be ready for delivery to one of your customers.

This is the basics of what's going on when you compile binary files to source code, download dependencies to third-party libraries down from the internet, stamp a version.txt text file into your artifact, and put everything together in a folder that you convert to a zip file and upload to your artifact repository on your platform. Tomorrow when you want to deploy this version, you can retrieve it using your unique identification number and initiate a deployment to nodes on an environment with the bill of materials inside the box.

Deployment

The software industry interprets the words "deployment" and "release" differently, which is kind of frustrating because this is a source of friction in day-to-day communication when seemingly similar words may be interpreted very differently across organizations and individuals. Because we as an industry haven't yet settled on a vocabulary that we can refer to, I want to state clearly what I believe constitutes a "deployment" and what constitutes a "release." Once you've heard my opinion on this topic, you should compare it to books and other content you will encounter elsewhere and make up your own mind.

In my interpretation, a deployment is the act of changing state on nodes in your infrastructure, like changing a file with a newer version of the same file. The experience of the end user does not change though.

Release

A release, however, is activating a change in functionality for an end user.

In reality, lots of automatic deployments are implemented in ways that bundle deployment and release activities together, meaning that a deployment is also de facto a release. Many teams don't use the opportunity to separate deployments from releasing a feature using concepts like feature flags[11] or similar approaches. There are lots of patterns and frameworks available allowing you to separate deployments from releases in different technologies with the benefit of being able to deploy changes in your systems without necessarily affecting your end-user experience at all. In terms of automation, it is important to distinguish between the two since you may fully automate deployments to an environment and also to your LIVE environment. A release would in this case be a decision, most often made by a human being who has evaluated and approved the quality of the preceding deployment that has taken place earlier.

Many companies never get as far as achieving this level of automation, and it is by no means the right solution to aim for in all cases. The price to pay for fully optimizing your deployments to LIVE environments may not be worth paying, considering the implications on your technological

---

11 Pete Hodgson, "Feature Toggles (aka Feature Flags)," MartinFowler.com, October 9, 2017.
  https://martinfowler.com/articles/feature-toggles.html

setup and your organization in general to be able to honor requirements like that. If 100% automated software delivery is the end goal, you should know that the last 5% of an automation effort may cost more time and more money to achieve than the previous 95%—and most companies don't need those last 5% to compete in the markets they are engaged in.

Pipeline

A pipeline is the sequence of activities you execute deterministically, one activity at a time, in the form of a workflow configured in your Continuous Integration and Delivery platform. You call the combined set of activities for building an artifact for your "build pipeline" and the set of activities needed to deploy an artifact for your "deployment" or "release" pipeline.

Different vendors apply different meanings to the words "deployment" and "release," so across technologies you may experience similar features being named differently.

Convert Your Existing Processes to Generic Build and Deployment Activities

Let's revisit Team Fun Dog's initial swimline diagram once more.

What the developer does in activity 2A and 2B, when he copies files into a folder located on the TEST webserver, is essentially build a rudimentary artifact. The developer has an intuitive understanding that it makes sense to build a package with changes to files, so in effect this has become the process depicted in Figure 16. The outcome in the example above is a folder with files that, based on the description of an artifact, may be defined as such. The quality and trustworthiness of its content is mediocre at best though. For starters, it has been assembled by a human being, the contents have likely not been tested and validated, and no versioning scheme exists, based on what we can see in this diagram. It doesn't change the fact that conceptually we have an artifact in the form of a folder to be used for deploying changes to an environment. That's what's being assembled in 2A and 2B.

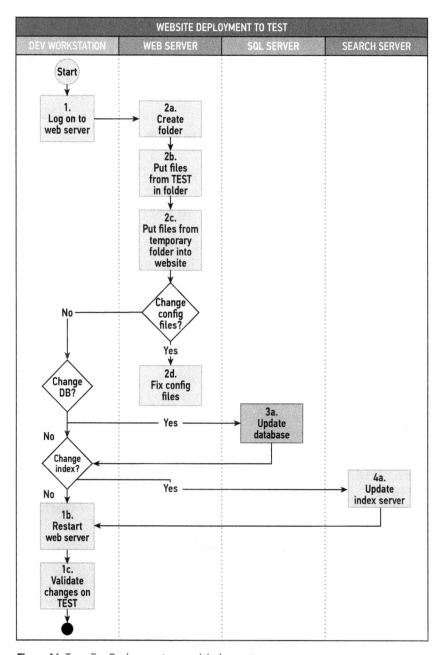

Figure 16. Team Fun Dog's current, manual deployment process.

If the definition of an artifact is that it should contain all knowledge necessary to deploy changes to an environment, you may have noticed already that Team Fun Dog has issues in the workflow outlined in the diagram. After changing state on the environment in activity 2C, they would have to open files manually in the root of the website folder and change things like settings. It is a severe shortcoming that induces the risk of variance to how many ways they can execute any given activity. The junior developer or newly assigned senior programmer might open and edit the wrong file, to name one example. Or you may open the right file but make a spelling mistake because you didn't know that configuration setting values are case sensitive. Rest assured, the sources of errors are countless when humans are involved.

When building an artifact containing all necessary knowledge in your build pipeline, you must do this in a way that all necessary work for building and assembling configuration files for each environment in question is an activity in your build pipeline. This will secure correct assembly at all times, and it allows for static testing of your configuration files afterward at the time of build yet prior to completing it. You may get a warning and the opportunity to fix a spelling mistake or the like before deploying your software to an environment.

There are some unfortunate processes right now that we need to revisit as part of automating Team Fun Dog's current release process. First we need to restructure their current flow so we end up with two flows: one for building their artifact and one that deploys changes to their environments.

The first flow, the build pipeline, contains the following generic steps (Figure 17).

Subsequently, when deploying software, they will need to engage in the following, generic activities (Figure 18).

All activities must complete successfully before the next one begins. In other words, the entry criteria for an activity to begin is that the previous one completed successfully. That means that if you aren't able to compile binary files based on source code because you checked in an error to your version control system, the remaining activities shouldn't be allowed to execute. The entire pipeline should fail and instantly report back that something went haywire.

Figure 17. Generic build pipeline.

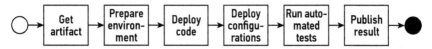

Figure 18. Generic deployment pipeline.

The activities described in Figures 17 and 18 describe a technology-agnostic series of activity umbrellas that you may implement in widely different ways across technologies and platforms. For that reason you may design a series of activities during an implementation phase all related to different ways of , for example, testing your artifact prior to it being published to the Artifact repository. The "Run tests" activity in one pipeline may consist of running unit tests, performing static code analysis, verifying the contents of configuration files, plus much more. Another build pipeline for a different product may not have any tests at all, so you only validate things like a file count or the existence of specific files and let that be it for now. All activities are automated audits validating quality and state, but the way they are implemented may vary greatly. What matters is that across all software builds, these steps are the ones you will need to consider implementing regardless of what you are building and subsequently deploying to nodes in your infrastructure.

You may benefit from searching for "Infrastructure as Code (IaC)" to look for books and content on Continuous Integration and Delivery best practices. This will get you started if you need some guidance for how to deterministically configure infrastructure reducing unknowns and variations of, say, installed software versions on servers such as build servers or application servers in your infrastructure.

The build itself should run on dedicated hardware of some sorts. A developer workstation isn't good enough. Don't even think about making shortcuts like that unless you're just playing around and doing proofs of concept. It is beyond the scope of this book to describe all the advantages of having dedicated resources for building your software, so I'll only urge you to do what's necessary to avoid depending on developer workstation machines for building your software. There is a low upper limit to how trustworthy the contents of an artifact can ever be when you don't know the starting point of how the software was created—which you don't, because developer workstations get updated continuously with patches, programs, and stuff downloaded from untrustworthy places on the internet. Don't shy away from using things like hosted build agents in the cloud to get started. They aren't free of charge, but the up-front cost is still cheap compared to designing for failure by not separating concerns properly in your build pipeline.

Create a Version Number

A version number is metadata for your artifact containing a unique identification signature. There are many schemes and ways of creating version numbers.[12] I subscribe to semantic versioning, which you probably already know. It looks like "version 1.5.23," where the number 1 is the major version number, the 5 is the minor version, and 23 is the patch number. I would say that semantic versioning is the default industry standard that most software vendors apply for products and services that they deliver to their customers.

Semantic versioning is used when others may want to consume software that you have produced, creating a dependency between their software and yours. The primary use case is application programming interfaces (API), but there are good reasons for versioning all digital artifacts that you and your team produce. If you're an advertising agency producing and delivering graphical material such as movie clips to a customer's LinkedIn channel, why

---

12 "Software Versioning," Wikipedia, last edit July 4, 2020. https://en.wikipedia.org/wiki/Software_versioning

not version your deliveries in a format that clearly communicates whether or not the movie clip is ready to be released? This way you can train your customers in the notions of semantic versioning so they can validate whether or not a file called "Linkedin_campaign_v0.2020.8.31.mpg" is the right one to release to production. With a little training, it will become instantly obvious that this isn't the file you would want to upload to LinkedIn since the 0 in the major version position tells us that this version shouldn't be considered production ready yet.[13]

If you raise the level of abstraction a bit, a delivery of a feature on a website or in an app, as compared to the delivery of an Adobe Photoshop file, are just digital deliveries to a customer. There are lots of valid reasons for using semantic versioning, which enables you to communicate what others may expect in terms of quality and content. You can reap all the benefits of doing so, like deciding on a naming convention than includes a version number or similar. The advantages of doing it consistently are hard to underestimate.

I always recommend that teams use semantic versioning regardless of the nature of their digital deliveries, especially if the team has yet to adopt a versioning scheme. Semantic versioning is well described, recognized, and supported across technologies. The importance of unambiguous naming is essential, which is why I push quite hard when I see that the root cause of troubles down the value stream is due to the lack or perceived unimportance of versioning. It is absolutely crucial when building software and digital artifacts that you don't make unnecessary assumptions that in turn create misunderstandings, which increases the support load and negatively affects the end user experience. Even though a version number in principle doesn't have to be more than just a number that increments between deliveries so others can handle basic dependency management in your systems, there are only good reasons for investing a bit of effort here. The benefit is improving traceability and quality in the delivery experience itself using a well-documented format. Systems and humans alike will only need to look at your version number to extract knowledge about the current state of the delivery rather than spending time and brain cycles inventing and maintaining your own proprietary specification.

---

13 "Semantic Versioning 2.0.0," Semver.org, accessed July 15, 2020. https://semver.org/#spec-item-4

Part of the version number may potentially be given, such as the major version. Microsoft developers alone don't get to decide if the next major version of .NET should be named 5.0, 6.0, or 7.0; there's a whole marketing division that needs to be involved too. Your control span as a team over the version number depends on the strictness of your release management processes of the digital deliveries that you and your team are responsible for. If you have no dependencies, you might want to think outside the box and investigate how you could create a version number that expresses metadata that may come in handy at a later time.

Let's take the example below, where you have a perfectly valid version number you could put into a version.txt file:

```
2020.8.31.2043
```

The human eye will be able to recognize this as a date, namely the 31st of August 2020 at 8:43 p.m. You would know exactly when this software was built. It is perfectly fine to stop here.

But we can do even more with semantic versioning. Let's look at another version number like the one below:

```
2020.8.31-beta.ifhw3m2g
```

The version number described above, where the patch number contains a dash, will according to the specification be considered a prerelease, meaning that it's not ready for production yet. When you match your tree of dependencies against that packages available to you, the latter version number will have lower precedence than the seemingly similar 2020.8.31.[14] This means that you can communicate in quite exhaustive detail about what the recipients of your deliveries can assume regarding overall end quality of a specific delivery simply by encoding this information into a publicly available format visible to your end users.

One snag, though, is the fact that you may experience difficulties using specific technologies if you use anything besides numbers in your semantic version number. Microsoft.NET is renowned for shortcomings in that

---

14 "Semantic Versioning 2.0.0." https://semver.org/#spec-item-9

regard. You cannot stamp anything but numbers into the metadata of binary files when you compile source code using MSBuild, which in my belief is due to backward compatibility and likely the legacy of a very unfortunate strategic decision many years ago. It doesn't change the fact that the specification of semantic versioning allows for a series of opportunities to pass on information about the level of quality for any given delivery to your consumers and end users. You will have to adapt to the constraints forced upon you by the choices in technology (as in the example with Microsoft.NET), but I urge you and your team to consider using the version number to add knowledge about things like the time of the software build or whatever you find will be a value-add in your context.

> One of the first things I always look for when assessing DevOps automation in companies is whether or not the version number of a delivery is available for end users. The existence or absence of a version number in digital deliveries indicates to me whether or not a team has a culture of frontline focus and measure of quality from an end user's perspective!

Version your software—there can be no other conclusion. Apart from that, you ought to make it visible to your end users which version of the software they're using, even if you're not forced to do so. Sometimes you are forced to provide a version number as part of your upload process, like when uploading apps to an app store. Far too many teams in my opinion do not understand how much users love it when they have information like that available to them. It's heartbreaking to an extent, because if you make a website where users from time to time experience incidents and hence file a bug reports, you will make both their life and the life of the second-level supporter in charge of debugging the nature of this error a whole lot easier if the user creating the support ticket could look up the version number, maybe in the footer of the website. The developer on your team would have a much easier job going forward if the version number was part of the original ticket, especially when working on highly dynamic environments where updates happen frequently or bugs are fixed continuously, such as after a larger release of new functionality.

This is just one single use case; there are lots of other situations where the perceived quality of your delivery is much less about whether or not a feature's functionality meets the expectations of your end users. The perception of quality is often as much about how good you and your team are at building knowledge and metadata into your deliveries and making them available to stakeholders. Taking this small step means these stakeholders will feel helped and catered to when they experience times of trouble and mishaps.

The examples of unambiguity are numerous. Versioning your deliveries removes the need for clarification with which exact delivery a customer experienced an incident or other means of poor quality in the end product. I would recommend that you always decide in favor of versioning and implement that strategy for all your deliveries as part of the automation initiative you've begun.

Build Your Artifact

Once you have devised a scheme for versioning your software, you should download, build, or in other ways realize the files necessary to deploy the changes you wish to perform on your nodes.

If you work on technologies that interpret source files during runtime, such as server-side technologies like PHP or client-side technologies such as JavaScript, your "build" process may consist of retrieving source code from your version control system and cherry-picking the files you need to be part of the resulting artifact. If you work on technologies that compile binary files from source code, you will need to download the source code from version control, compile them, and then copy the binary files to your artifact folder.

A generic build pipeline activity called "build your artifact" therefore covers a long series of activities, just like the generic "run tests" activity may cover running automated tests, performing static code analysis, etc. We can take minification of JavaScript files as an example of a known technique for optimizing their size, thus lowering download times and improving browser performance. The tradeoff is that an optimized file is not human readable because things like line breaks are omitted, variable names are shortened, etc.

You will for that reason need two versions of the same file that are similar in functionality: one that you can read and use when debugging incidents and outages, and one that is optimized for performance and meant to be used in your LIVE environment. When assembling your artifact, you will generate all versions of the files that you need. Your artifact should, as mentioned earlier, contain all the files you will ever need across all environments, so even though the optimized version of your JavaScript file is only necessary for your LIVE environment, this is the time and the place to construct it.

When both files are available in your artifact, you will at the time of deployment choose which of the files you wish to deploy in your environments. If you have a development environment that you have decided to configure for ease of debugging and triangulation of errors, it would be beneficial to deploy the version from your source control as-is while in your TEST and LIVE environments, you want to test the compressed file, since compression and optimization is of greater importance to browser performance and thus the end user's experience.

Some teams design their pipelines so that they are not testing the very same artifact that ends up being deployed to the LIVE environment (see Figure 19). There are lots of variations in this pattern, and the reasoning behind them often resides in flawed version branching strategies. In call cases, this is an antipattern you do not want to implement regardless of context. The primary issue is that if you build artifacts A1 and A2 based on the same revision of your source code but at different timestamps on a timeline, you're not guaranteed that the contents of your artifacts are the same just because the artifacts were generated on the basis of the same source code revision. What happens if a team member changed the build pipeline between the time of building A1 and A2? What if A1 was built on one build server and A2 on a different one, which may not be configured like its predecessor? My guess is that you and your team will one day end up spending a lot of time debugging ghosts and weird behaviors only allowed to happen because tiny differences between artifacts A1 and A2 slipped through your automatic tests and other means of quality assurance. For that reason, the artifact should contain everything for all environments, and it should always be the same artifact you retrieve and deploy on nodes in your infrastructure.

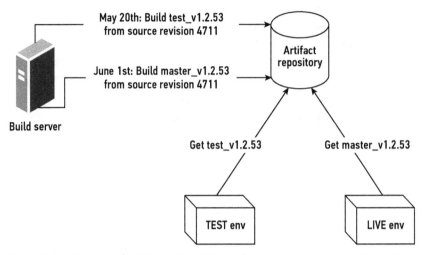

Figure 19. An antipattern of building dedicated artifacts for each environment. Avoid this pattern —always deploy the same artifact to all environments differing only in terms of configuring a deployment with settings specific to the environment in question.

Place everything in a folder structure defined by you: the outcome of a compilation of source code, your compression and optimization of files, your selection of files downloaded from version control, your generated version.txt file, script files, plus all other remaining activities. Once you have completed all activities related to building your artifact, you will have all files and folders gathered in a folder structure that you will later be able to download and extract on nodes in your infrastructure.

Size Matters

Size does matter—no pun intended. If your website amounts to a size of, say, one gigabyte of files and blobs of data because you've done as I told you and put everything into the same artifact, you'll find that it will soon take an unacceptable period of time to juggle your large artifact when transporting it from one datacenter to another across dedicated, swamped internet connections. It also takes a long time to build your artifacts and convert them to a compressed file, just as it takes a long time to unzip them once you've downloaded an artifact on a deployment target node.

The perception of performance and, in turn, the perception of quality of your build and deployment pipelines deteriorates proportionally with the wait time of each operation in your pipelines. You will never achieve a state of flow mentally if you have to wait fifteen or twenty minutes for feedback every time you commit a small change to your source repository. It also requires a lot of I/O on build server hard drives and a lot of disk space on hard drives on all your nodes when, during debugging, you deploy changes multiple times a day for a period of perhaps several weeks. Unless you have mitigated the risk of disk space starvation by setting up strict retention rules for the folders where artifacts are downloaded on nodes in your system, so that old ones are quickly wiped out on a schedule, you may find that artifacts have consumed all available disk space on deployment targets. The problems with handling large artifacts are numerous, and they will ripple through your infrastructure if you're unaware of the implicit consequences of building and handling artifacts of a certain size in your systems.

A large artifact of several hundreds of megabytes equals technical debt and should be mitigated as such within the frames and processes of how you usually prioritize paying down technical debt in your portfolio of systems.

It's easy for me to just say "Divide and conquer! Split up your large artifact into smaller lumps and deploy what makes sense in context." It's easy to say but harder to do right, especially if you don't pay respect to the complexities involved when managing lots of smaller, interdependent artifacts interfacing with each other. Small sizes reduce footprints in your systems and will help you improve both build and deployment lead times, but the tradeoff here is the fact that you and your team will need to have a strategy for dependency management, meaning that you will no longer have one pipeline building the entirety of your artifact as one, single set of activities. You will likely find yourselves forced to implement several build pipelines, each capable of producing a subset of the bill of materials needed to fully deploy your digital delivery from scratch.

Complexity and the resulting risk of something going wrong will increase with the number of dependencies you build in between different versions of artifacts. Choose wisely where to cut when you decide to chop up your big, monolithic artifact into smaller ones. It can be challenging, but it's far from impossible. You will likely at some point be required to build more than one pipeline as you invest more and more into automating several interdependent products in your systems.

An example of what a large artifact may contain is a legacy website with a lot of image files embedded in a subfolder, say 200 megabytes. Without the images, the website doesn't work, but the image files themselves never change. Only new ones are appended from time to time. You probably wouldn't build a website from scratch like this today, but for the sake of the argument, let's just proceed with the example. There are lots of legacy solutions out there that resemble this setup. The manual process of deploying changes to source code has worked fine because they would only ever change source code files and not the image files. But now the deployment has been automated, and all image files have been copied into the artifact. Whenever new image files need to be pushed to the LIVE environment, images are simply uploaded to the version control system in the correct folder. A new artifact with the new image is built and then deployed. This build pipeline effectively builds an artifact of several hundreds of megabytes every time you change a comma in a JavaScript file.

This is not a good place to be. The feedback cycle grinds to a halt whenever a developer working on JavaScript changes has to wait for build servers to download hundreds of image files, copy hundreds of megabytes to an artifact drop folder, zip those same hundreds of megabytes, and push hundreds of megabytes to an artifact repository—just because he changed a few characters in an unrelated JavaScript file.

You want to achieve two things: You want to have everything in source control, and you want to be able to quickly deploy changes to your source code without paying the penalty of handling hundreds of megabytes of unrelated files every time. That's a fair approach, and they are by no means mutually exclusive.

Source code, to the extent of my knowledge, consists of flat text files in just about every programming language. They are never the things taking up much space, even though they have the majority of changes happening to them. One way of solving both use cases is to design two build pipelines. Place all files that change very often in one swimlane and put everything else in another one. In this case, you would put all your source code in one build pipeline and have all your images in another one. This way, you would construct two artifacts: one with your source code and one with all your image files, which you would be able to build independently from each other.

During a deployment you would need to download both artifacts to your nodes. That may potentially incur a penalty if you're forced to download both the small one and the big one, even though only the small one had any changes to it. The solution would be to investigate a means of caching your artifact download on nodes, which should be a configurable feature built in to your Continuous Integration and Delivery platform. This way, the platform would determine that the bigger artifact is present and hasn't changed since the last time, so only the smaller one should be downloaded and extracted.

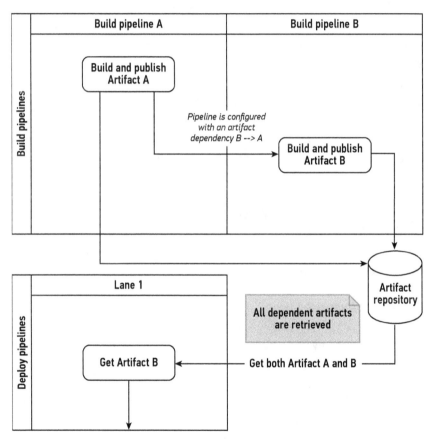

Figure 20. One approach of speeding up build and deployment cycletime when the root cause is the size of the artifact itself is to have one build pipeline capable of quickly building sourcecode you change very often and one building an artifact containing the subset of the entire delivery that rarely changes.

You may also, during the deployment itself, configure your way out of not copying anything into the image folder of your website if you have already successfully completed a download and deployment of the image artifact before. Or just copy new or changed image files from the artifact download folder into the website. There are lots of patterns available to you to increase speed of delivery and execution of tasks. This pattern of segregating work based on size of produced output is just one example. The basic principle in any case is to separate big deliveries into smaller ones that may be built and deployed independently.

Should you save all image files in a self-contained build and deploy pipeline while having another pipeline deploying output generated from source code? That's one viable option. You may decide to migrate all your images into a content management system (CMS) tailored to handle and version image files instead of placing large files in a subfolder of your website. That would eliminate the need for different pipelines and perhaps leverage the benefits of an improved lifecycle management for images and other content on your website by using built-in features of your CMS system.

Regardless of the strategy, the way of dealing with large artifacts is one of divide and conquer. Utilize dependency management of some sort so you won't have to build and persist hundreds of megabytes every time you need to change a spelling error in a JavaScript file. Keep an eye on build and deploy timelines as you progress. It is the size of the artifact in combination with the structure of your build and deploy pipelines that more than anything else predicts performance and perceived quality of your automation efforts.

Execute Automated Tests

"TEST!"—that was the single word written on the whiteboard at the office of Just-Eat.com, my first employer after I graduated as a programmer back in 2003. I was employed as a software developer in what was then a small startup with ten or twelve employees, and I had a track record of recurring, horrible mistakes in the code I produced for our web shop, which in one instance caused bad things to happen when customers tried to add items to the shopping basket. Due to immature technology stacks and lack of experience, I learned that a comma and a period in ASP.Classic, the programming language of choice at that time, could be interpreted differently if the language setting on the server was set to English instead of default Danish. The consequences were that a pizza wouldn't cost $8.50 USD as expected, but a mere $8,500 once the shopping basket was converted to an order in our ERP system.

I also took care of telephone support at that time during office hours, so fortunately there was a tight feedback loop that allowed me to fix some of the errors quickly. But from time to time, some of them slipped through and went unnoticed all day, until either restaurants or angry customers called our evening staff, offloading their wrath and perceptions on our (read: my) ability to deliver up to standards.

The cofounder and CEO at the time, Jesper Buch, would occasionally take me and my more senior colleague aside and try to explain to us how frustrating it was for him to be working eighty-five hours a week without pay just to experience a mailbox full of unhappy customers because we introduced grave errors over and over again. At that time, technologies were less than mature, and online shopping was a concept that only a minority of internet users had ever tried firsthand. We built the web shop solution and back office solutions from the bottom up, while my boss and his friends, together with a couple of other fulltime employees such as myself, tried to teach the residents of Denmark how awesome it was to order pizza on the internet through our website rather than calling the pizza shop directly using a telephone.

With lots of enthusiasm and very little experience, I was responsible for building, scaling, and maintaining all technical parts of a web shop together with a colleague. It went well—the company still exists and is doing well at the time of this writing—but it wasn't due to the inner beauty of the software I wrote, I can tell you that! I didn't know anything about the entire craft called "software testing," didn't know the different types of tests you can make and how to structure your testing efforts. We encountered at regular intervals massive problems due to the quality of my work, especially during times of exponential growth. But I had no clue about the basics of software quality assurance, and it did cost a lot of money in terms of lost orders and unhappy customers. Even worse, it took its toll on my belief in my own abilities to fulfill my position. I didn't know at that time how complex it really is to ensure the quality of a software delivery so that you can really have confidence in it without lots of disclaimers and preconditions written in small letters somewhere.

I cannot stress this enough: You and your team will not experience success with Continuous Integration and Deployment in the long run if you do not have basic competences in writing automated tests of your deliveries!

As a software developer, I have since then spent countless hours learning the ins and outs of automated testing and learned how to write both code and the surrounding automation pipelines. I can now ensure that you as a team and organization achieve the quality necessary to satisfy demands and expectations around operational stability and predictable numbers of support tickets after a release. It's a steep learning curve, but if your craft is software engineering, you will need to learn how to write automated tests. You will not become successful in your automation efforts without being able to instruct software to take over parts of the quality assurance of the digital deliveries you wish to promote to your end users. As a leader, you increase the likelihood of promotions quite a lot if your employees are able to deliver high-quality digital deliveries on time and within budget. This is where research in this matter is very clear: automated tests are an expense up front, but they are an investment that correlates with higher quality in the resulting digital delivery, especially when the test suite focuses on regression tests run multiple time. The investment will pay for itself over and over again once implemented.[15]

If you do not have the foundational competences in writing a suite of automated tests for your digital deliveries today, I would recommend prioritizing the investment in education on these topics before you focus fully on automating what you now do manually. You and your team have to be in a *very* bad place indeed if the ROI for automating build and deployment of your deliveries supersedes the advantages of being able to write automated tests of the code you're producing. I would take the stand that if you were only allowed to look at one single factor when evaluating whether or not a

---

15 Divya Kumar and Krishn K. Mishra, "The Impacts of Test Automation on Software's Cost, Quality, and Time to Market," *Procedia Computer Science*, December 31, 2016. https://www.researchgate.net/publication/300080121_The_Impacts_of_Test_Automation_on_Software's_Cost_Quality_and_Time_to_Market

software developer is competent and capable of adding value, I would look at his or her ability to write automated tests. The ability to write automated tests is in my opinion the single largest contributor in predicting whether or not a software developer is able to deliver high quality at a predictable pace with trustworthy estimates.

It is beyond scope of this book to teach you and your team how to become proficient in writing automated tests in your context. It depends on the technological constraints of existing competences, your background, the amount of technical debt and so on. Years ago it could be troublesome to write automated tests for front-end technologies such as JavaScript, but the technologies and tooling available have matured a lot during the last six or seven years. We have seen progress on that agenda in just about every technology that I am aware of. I believe that at the time of writing, it is possible to find frameworks for writing automated tests today in just about any technology used by companies. There may be severe constraints and limitations, but the need for proper tooling to write automated tests against code written in various code languages is there, which is why companies provide means of writing those tests.

The amount of quality material for online learning is also profound based on where we were at just a few years ago. You can learn the basics by taking online courses and reading books. You should do that regardless to live up to basic expectations in terms of skillset and competences, I believe. The principles are uniform, but it would none the less be beneficial for you and your team to ally with an expert in the field introducing basic principles of test-driven development (TDD) when engaging in an automation initiative in your context. You should be aware of and humble about the fact that context determines whether or not it is a good idea to automate build and deployment of software component A rather than component B, even though the principles of automation are the same for both A and B. If A is littered with technical debt and B is not, but the overall end-user quality of component B has less strategic importance to the company, the image gets blurred quickly in terms of right and wrong. If you have a need and a concrete use case but haven't done this before, you should find somebody who can apply the basic principles into your context and help you get started in the right direction doing automated testing of your digital deliveries.

Test-driven development (TDD) is a software programming discipline where you, in combination with the code implementing business requirements, write a series of tests that validate requirements as the developer implements them. These can then can be bootstrapped to run automatically for regression test purposes after the implementation of a given feature is completed.

The V-model

Even though I can't provide an implementation recipe for what you should do to ramp up, I will provide you with a plan of attack that you can use to create a strategy for automated testing. This blueprint for creating your deployment pipelines should improve stability and increase the speed and timeliness of crucial feedback that's central to getting value out of the effort invested in automation projects.

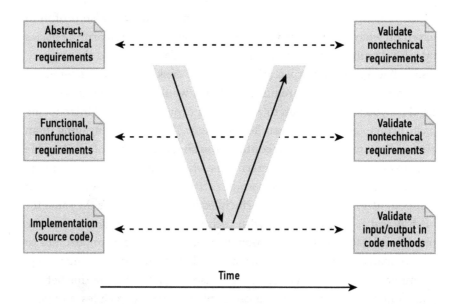

Figure 21. V-model, where you plan your test phase in parallel with breaking down high-level business requirements into smaller, concrete deliveries such as userstories.

Most software testers will nod their heads when asked if they're acquainted with the V-model. It visualizes how you may structure your testing efforts from the very beginning of a project where you're in the progress of defining scope for high-level requirements but haven't yet broken down work into a size that a delivery team can execute and deliver upon.

At the top left of the V-model, you have a concept or business demand regarding a digital deliverance. You may say that the level of abstraction is highest at the top. The bottom, on the other hand, is concrete—in the case of software, these are your concrete implementations; your source code, for instance, is located at the very bottom of the V-model. The right leg of the V-model illustrates an ongoing test effort that runs in parallel with work commenced on the left-hand side. This model breaks down high-level, abstract business demand into more tangible, concrete lumps of work. The left side of the V-model is effort spent on delivering software; the right side of the V-model is work spent on planning and executing the series of tests ensuring that scope and quality requirements are met before deliveries are released to end users.

Let's say you have an abstract, nonfunctional demand such as "Performance should be on par with the existing solution." We could write this demand on a note and stick it to the top-left side of a large V on a whiteboard. Ideally, at some point before you and your team start writing any code, you would devise an acceptance test for this abstract demand. You could write, "Is performance acceptable prior to delivery?" Attach this note to the top right of the V on the whiteboard.

Once you break down the abstract demand of ensuring good performance, you may decide to formalize what good performance looks like. You eventually decide the following: "API response time should be < 200 milliseconds." This demand should be followed up with a similar suite of tests ensuring that API response times are actually below 200 milliseconds. Those two sticky notes should be put on the same horizontal level in our V on the whiteboard. This way testers can work with stakeholders and the team on all abstraction levels to plan a suite of tests that validate requirements, functional as well as nonfunctional. You shouldn't first engage in the testing phase at the end of an implementation phase. You wouldn't know what to test if you haven't had discussions early on with stakeholders about the expected outcome of

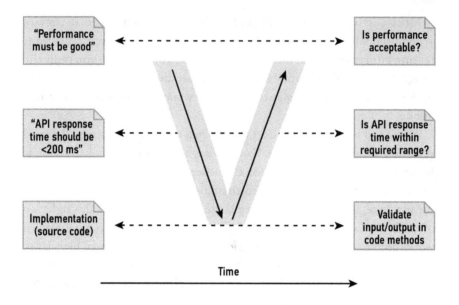

Figure 22. Concrete example on how to design tests based on highlevel requirements being implemented.

a certain feature. It is during the requirement breakdown that you have the best conditions for planning and later on executing an adequate suite of tests ensuring functional and nonfunctional requirements for your overall software delivery.

Not all organizations understand what a tester is doing and why software developers aren't good at testing their own deliveries as well as somebody who makes a living by being outstanding at testing all kinds of software deliveries. Test managers use the V-model as a tool to communicate to their stakeholders that the work of testing a delivery must commence the second you decide to scope a project and focus on implementing the first delivery.

Quality is an attitude—and hiring a test resource requires that a stakeholder above you with budget and decision-making power has that attitude. If you find yourselves in situations where you're end-to-end responsible for testing all deliveries from your team without proper competences and knowledge, the V-model may help enabling you to communicate in the context of a simple and validated model.

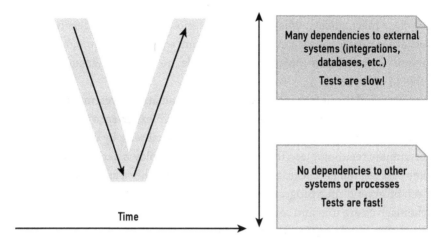

Figure 23. Reduce variance by having all environments retrieve the artifact from the same location always.

Speed Is Key to Success

There is a somewhat subtle point that I feel doesn't get the attention it deserves when you discuss the V-model in relation with Continuous Integration and Delivery. The point is that the amount of dependencies increases the farther away from the source code and concrete implementation you get. I sometimes show my own version of the V-model, like the one below, when I facilitate, say, unit test workshops for teams an organizations, because there are certain types of automated tests you should not want to run by default as part of your build pipeline.

All automated tests across technologies share a set of characteristics that makes the individual test either a good or a poor fit for being executed as part of a build pipeline. The primary characteristic is speed. Only run tests in your build pipeline that can be executed in just a matter of two-digit milliseconds—those tests are called "unit tests," whereas tests with more dependencies will typically run for far longer, such as several seconds or more. There are lots of nuances and a constant naming game for what constitutes different types of automated tests, but for the sake of simplicity, I will settle for two categories in this book: unit tests and integration tests. The difference is that a unit test can execute from start to finish in a matter of few milliseconds, in effect having no dependencies to external processes or state mechanisms, whereas an integration test—due to dependencies on other processes such as database instances, API endpoints, etc.—will require cycle times in the matter of seconds, no matter how much you are able to optimize it.

I have had the unfortunate experience of evaluating investments in automation projects where individuals without much experience in test-driven development have engaged in automating manual tests by using Selenium or similar technologies. These are all able to replay a series of keyboard key presses, mouse clicks, and similar events in either a smartphone app, a web browser or in a program installed on your computer. The technologies themselves are fine and serve a well-defined purpose for those who know what they are doing. But I will dare to conclude that you don't if you determine that Selenium is the right place to start when diving headfirst into a larger initiative for automating manual testing without experience with automated testing whatsoever.

The header of this section states that "speed is key." You wish to focus your first efforts where the speed of test execution is a matter of a few milliseconds. You won't be able to get that kind of speed in your automated tests unless you avoid dependencies all together. The only place that you don't have dependencies to external systems is in the very bottom of the V-model, close to your source code. It is the craftmanship of taming tests at this level that you will need to learn first.

Before you will be able to execute an automated test using something like Selenium, you will need to have the program under test installed first—or a website running that hosts the version of the code you wish to execute and validate your test suite against. This application or website should be able to access all necessary data in an acceptable level of quality form, such as a database or similar data storage facility. You'll also want to know exactly the contents of all raw data being retrieved by the system under test before you execute any tests at all. Assumption is the mother of all evil, even in the field of automated testing. You need to know your starting point, which is a complexity that requires skills, work, and experience for success, even for highly experienced and well-founded teams.

There's no final demand stating that you need to know each and every bit of data up front, but I see lots of teams struggle and have problems with consistently executing test suites of integration tests because they execute their tests against a foundation of data that aren't known to them. It may be a backup of a production database, which makes good sense in a lot of scenarios. But if in your test setup you assume that a user with the email john@doe.com, for example, always exists and the user deactivates his account on your website ,meaning that the entire user account with all history gets deleted thirty days later to adhere to GDPR regulations, you will experience a lot of test cases failing for no apparent reason without having changed anything.

You and your team may work around issues like that, but only as long as you never allow "accepted" unknowns in your test setup to cause periodic false positives or false negatives. As you scale and build on top of a test suite that cannot be trusted a hundred percent, you will put the entire test suite at risk and pay opportunity costs without harvesting the gains in the form of higher quality end product. You risk losing faith in the test suite you've built. This loss of faith will happen at the expense of testing efforts and other quality initiatives you could have made had you chosen a different strategy of execution from the beginning. The team may also be reluctant to decide in favor of removing flawed tests or ditching a large test suite entirely that they have put a lot of effort and perhaps political capital into. This hesitation may happen even though it has been a well-established fact for months or even years that they will never be able to refactor the tests into a valuable asset anyway, in effect living out a textbook case of "sunk cost fallacy." It is harder than you may think to write automated tests due to a steep and unforgiving learning curve. It requires quite a lot of experience and is difficult to get right, especially if you write tests for systems riddled with legacy and technical debt. You will need to have tried and failed a few times before you experience long-term success in this regard. Rest assured, there are lots of developers suffering from the Dunning-Kruger effect[16] in regard to proficiency of writing automated tests.

---

16 "Dunning-Kruger Effect," Wikipedia, last edit July 7, 2020. https://en.wikipedia.org/wiki/Dunning%E2%80%93Kruger_effect

It can be hard for newcomers to TDD to understand that it isn't apparent to the untrained eye when a unit test is indeed a unit test according to the definition outlined above. A fast-executing test suitable to be described as a unit test changes characteristics when you introduce a dependency that slows down execution time. You may introduce a database lookup to a unit test to retrieve some value that you will verify against the output of an algorithm, or you may call an API endpoint for one reason or the other. Very well, what you have is no longer a unit test with totally self-contained datasets and without dependencies to external systems. To the untrained eye, your unit test is still exactly that, but in reality it is not anymore. It has dependencies and runs slowly, but this isn't necessarily something you can see by looking at the test code itself. You will need to time your test and perhaps profile it, too, to get more data about what's going on under the hood.

Get good at writing "white-box" unit tests first, meaning tests without dependencies and that are self-contained in terms of all data needed. Don't rely on database instances, APIs, or similar means of data storage. Once you have learned and are competent in using mocks, stubs, Inversion of Control (IoC), and perhaps a few code design patterns as well and can apply those principles in a meaningful way in your legacy codebase—only then would I suggest that you begin working your way up the V-model. Then you can write automated tests in the form of black-box tests that invoke the interface of a component or a service getting an output without knowing (or caring) about how the service produced the output returned.

Once you become good at writing white-box tests, you will quickly adapt to writing black-box tests to test multiple components in combination with each other. Don't get me wrong: there's absolutely nothing wrong with writing integration tests. On the contrary, they are invaluable for testing all of your deliveries in combination with each other. It is very helpful to know that you have tests in place to, say, validate the length of a person's name and an integration test validating, from the end user's perspective, that you actually can still create a valid user in the database before you release anything. Just make it very clear to yourself that you don't start with the latter! Integration tests build on top of unit tests, not the other way around. You need competences in writing white-box tests before you move on to writing an integration test that, for instance, invokes an endpoint in your service with some fake person-like data checking that everything works as expected from a customer's point of view. In my experience, the tutorials and other material on how to write automated tests tend to focus on white-box testing without emphasizing enough that, even though you are able to write an automated test, it takes time and perseverance to become skilled at writing a well-performing, scalable suite of tests with low maintenance overhead in the years to come.

If you are working on or have inherited large test suites of slow-running integration tests, take this approach: recognize and value the fact that they are there for a reason. It is a problem that they are running slow, and it has consequences in the form of increased complexity and prolonged feedback cycle time. But that price is worth paying if the alternative is to delete the test altogether so you will no longer have any regression tests validating the scenario the test was originally written for.

You have options here. Rework integration tests and make them unit tests if you can. It may not make sense, and refactoring tests like that usually takes a lot of time, hence it is an expensive exercise to entertain. Another option is to not run all tests always, but maybe just run the slowest ones on a schedule. This has drawbacks, but it is absolutely a viable pragmatic approach that may work in your context. It's better to run 95% of all tests in 2 minutes than 100% of all tests in 30 minutes. Swimlane the slowest running tests in a separate test suite and run that suite on a schedule, e.g., once per day or at 9:00 a.m., 1:00 p.m., and 4:00 p.m. You don't get instant feedback, but you get quick feedback from 95% of the remaining test suite at the expense of getting feedback from the remaining 5% of tests potentially a few hours after a change was introduced to your source control system. You be the judge whether or not that would be a viable solution in your line of work.

There are other options. You may parallelize running your tests. It increases complexity quite a lot, especially if the tests are slow running because there are lots of dependencies to external systems that may not scale to handle a doubling or quadrupling of endpoint requests just to name one bottleneck that may vouch against parallelization. It's worth investigating pros and cons nonetheless. Imagine that you have a test suite of, say, a thousand slow-running integration tests that you inherited from a legacy system that you're not in charge of maintaining. You could split

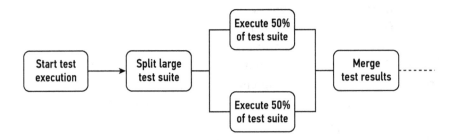

Figure 24. Parallel execution of a large test suite.

up the test suite into smaller portions and then let multiple processes run each test suite in parallel once you've built your artifact.

This approach is suitable only for platforms capable of running test suites in parallel out of the box on the technology stack you're using. I'll advise against you and your team ever writing code for parallelization if you can avoid it because the complexities cannot be underestimated and the gain is less noteworthy when you calculate cost in terms of how much support and bug fixing you take upon yourself and your team.

Parallelization is once choice out of many. By default, you should aim for a test suite running all tests every time you build your software. It's by far the easiest, and you avoid the risk of having your test feedback come in so late that you're way out of context when you get feedback that a test has failed. Now you need to task switch back into what you were doing at the time of the code check-in. The fallacies of task switching need not to be mentioned here; let's just say it hurts your ability to be the best corporate citizen that day if you repeatedly need to go back and fix errors or mishaps because it isn't possible to get feedback from your automatic regression testing while in the context of the problem you're trying to solve.

Make an effort to bring down the time from an introduction to a change in your versioning control system until all tests have executed successfully. There's a rule of thumb in the industry that it should never take more than ten minutes to execute a full build, including all tests. This timeframe is what you should aim for when you optimize time spent building artifacts. To get there, your test execution may be exactly what you should focus on optimizing to get cycle time down to an acceptable level.

Publish Artifact

The hardest part is over by now. The construction of your artifact and the quality installment of it is where the largest complexities are embedded when configuring a build pipeline. Your artifact is now ready and has been validated by automatic tests in one way or another. Now it should be put in stock and registered in your bookkeeping, allowing you to find and retrieve it at some later point in time.

The publishing itself is usually a simple operation on most platforms. You tell your platform the location of the folder on your hard drive that you have used to construct your artifact and put a nametag on your artifact. The platform will take over from here, compressing your bill of materials into a single file with a smaller footprint that is much easier to transport across the wire.

> The trustworthiness of your artifact must be absolute! The way of establishing the credibility of the contents of an artifact is to limit access to artifacts that have been built and versioned.

Some platforms give you the ability to put your artifacts on a network share where you as an administrator of the platform define access control. This is a perfectly valid approach, but if you do this, I strongly recommend that you restrict access to the network share so only the platform itself and the administrators of the platform have access to the network share. The reason is that IT technicians and software developers alike can be tempted to fix small errors in an artifact by editing the artifact directly in the artifact repository rather than fixing the root cause, meaning that the build pipeline or source code that is erroneous some way or the other should be fixed instead. If you enable write access to teams and individuals who are contributing artifacts to your system, you can be certain that somebody will eventually find out that there is an opportunity to edit the contents of a registered bill of material. That person will remember to use it in times of trouble when there is an outage and a team will pay the penalty for not having optimized a slow-running build, preventing them from quickly building and deploying a new version within a reasonable range of time.

Believe me, I've been a software developer for years, and know that where there's will and even the tiniest opening, there's a way!

It's best practice to name your artifact in a way so it contains at minimum the product name, also the version number if this is possible. The traceability of having a zip file with a name corresponding to the build that created the artifact is invaluable when debugging an incident occurring on a node that is not behaving as expected after a deployment. Some platforms have a convention for how your artifact should be named. If you have influence over the naming of artifacts and the platform allows you to parameterize (most do, I would expect), then you may name your artifact to the extent that the name itself describes the content, such as this:

```
Webshop_2.1.45.4283_20200405.zip
```

Just by looking at the file name, you can deduct without opening the artifact's version.txt file that the artifact contains your web shop, the version number is 2.1.45, and the time of the creating of your artifact was April 5, 2020.

It's a small thing to do, naming your artifacts consistently, but it is extremely useful in complex scenarios to have the version number and the time of creation available at your fingertips. The timestamps of the file itself that are embedded in the file format by the operating system aren't necessarily to be trusted. If the file was created fourteen days ago and only yesterday was copied onto the node that you're investigating, you would have no way of deducting from metadata on a file when it was created since you would only ever know that the file was created yesterday. You want the actual date of creation available to you, not just the date of the copy created on a node you're investigating.

Publish Result

Publication of your build result does not have to require more than what your platform offers out of the box. It may be enough to just log into your platform and check whether or not a build went well. You and your team do not need to do more for now.

However, as you progress and want more ROI from the time and money spent automating manual labor, you will need to raise the level of ambition a bit. An example could be that you might post a message to a Slack channel once a build has finished. If Slack is your primary means of communication, that may make sense. I would say, though, that you will get tired very soon of having your build pipelines pushing automatic messages about builds into your feed all the time. Information should stay relevant in context for it to be considered important. You do not in any way want to contribute to information overload, which obscures relevant information with unnecessary noise from various parts of a system blabbering on about events happening that you don't and shouldn't care much about because you're engaged in other projects for the time being.

Most systems offer features that allow users to turn on and off notifications on a per-project basis. This is a sound principle, because occasionally you want to know everything about anything on a particular project or pipeline because it has your interest for the rest of this week. In a few weeks, you will be assigned to other tasks, rendering the notifications from last week irrelevant. I celebrate self-service principles, where users and stakeholders are allowed control over what they get and how notifications are delivered to them. Stakeholders want to be advised, but it should be tailored to fit into their context, and there may be great variations. Which notifications matter differs from team to team, and more often than not, even from individual to individual on the same team.

Avoid notifications at all if everything went according to process. You automize and validate quality, and if everything is okay, why interrupt people just to say, "All okay"? Everybody involved is responsible for limiting white noise, so it's just as important to not communicate at all unless necessary, just as the receiving party should turn off his mobile phone and Outlook calendar, and select "Do Not Disturb" if he needs some time off the grid to concentrate.

If something did in fact not go as expected, then anybody who can fix an error in the build pipeline in question should be alerted immediately. This should be the default. You may have agreements in place in your team giving a "red" build priority. A red build in an otherwise healthy and well-performing build pipeline isn't bad—quite the opposite. It is an inadequacy in terms of not meeting the standards of acceptable quality— and you found this inadequacy, not your end users and customers! Job well done, good work on writing those tests, this is the exact second they provide ROI to you and your team. Let's fix this issue and make the build green—and get some cake too.

It is in times of trouble that you learn the most—if you are in a position able to learn from your mistakes. I believe that a team's response to unexpected red builds tell a lot about the maturity of a team and even more about how the organization supports and values a culture of learning.

First and foremost: Keep it simple, and make notifications a decision of the user receiving them. Opt-in solutions work best when you build and release software on different schedules and with shifting, strategic priorities.

....

Summary

Building your software is the first, crucial step. It is possible, when you engage in your first implementation of an automation project, to simply let the build itself become the first tangible delivery. You may set up an intermediate process of building your artifact automatically while deploying it manually until you have designed and implemented a deployment pipeline. It would be beneficial, though, to go through right away with the deployment process itself, as the friction is likely to be largest due to cross-organizational boundaries. The opportunities and gains from automation will be largest once you can demonstrate your abilities in the deployment of an artifact to an environment.

Not all steps are necessary to achieve quick wins. If you have a slow-running test suite that is impossible to run for various reasons, but you can create a build pipeline in a day or two without any tests run, this build pipeline will still be a value-add because you have automated the building of your bill of materials. You cannot guarantee the quality in terms of having run a set of automated tests against it, but you will never again experience copying the wrong files into the wrong folders, which will eventually happen when you ask humans to perform such tasks following a recipe in a wiki-article. Breaking down work into smaller deliverables, such as first building artifacts without running the test suite and then enable building and running tests in the same pipeline, would be a good plan of attack.

Part 4

How to Deploy Your Software

IN PART 3, we worked on building the bill of materials for all files and configurations needed to change state on nodes in your infrastructure. We also investigated pros and cons for how to publish the result of a build. Everything has been packaged and versioned into an artifact containing all knowledge necessary to change state. Now we're going to implement the activities to actually produce the changes you want to achieve.

It is at the moment of deploying a change to a node that the risk profile changes. The consequences of a breakdown in your pipeline are most often limited by the fact that the team itself has a problem they should solve. A red build in itself has no direct and immediate consequences for the end users or consumers of services. The pulse doesn't go up until a software delivery is handed over to the customers, because that's when you always ask: Will the delivery work as intended once it's put into production use?

A deployment is an operationally hazardous situation because you introduce variance into your systems, which may cause a worse end user experience while the changes take place. I'd like to walk you through how you should structure your deployment pipeline in a way that encourages quality in terms of an end user's need of operational stability but also takes the delivery experience itself into account in terms of robustness and transparency.

By now, you and your team should have implemented a build pipeline capable of producing a machine-generated, reproduceable, and versioned artifact available for download on nodes in your infrastructure.

It's the Environments, Stupid!

When you talk about preconditions for deploying software automatically in your environments, you will at some point need to actively decide where to deliver your digital delivery. A precondition for a satisfying end user experience builds on top of the inherent quality of environments used both to test the delivery and to host the delivery itself on the environment you have chosen, which should respond to requests made by your customers, i.e., your LIVE environment

It may be painful in many ways to automate a manual process. It may, for instance, be agonizing as a team and as a human being to be forced to change behavior and discard manual processes that you feel have served you well for years. It may be a source of uncertainty to be involuntarily committed to using technologies and problem domains that don't fit well with your current expertise and personal preferences. It may also be overwhelming for a team and an organization to be required think holistically about their surroundings. The complexities of understanding cause and effect of one's own systems may leave no bandwidth left to absorb complexities and abstractions not directly related to you and your team. A team might be fighting small and big fires every day, leaving no room for reflection or sound architectural decisions aimed at fixing root causes. I believe that the upfront cost of automation is proportionally equal to the amount of technical debt, the competences required to service that debt, and the wellbeing of the infrastructure on which software will eventually be deployed by a Continuous Integration and Delivery platform.

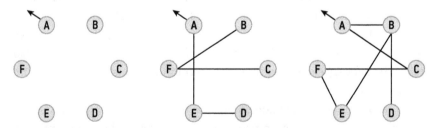

Figure 25. Dependencies increases the risk of unforeseen events in parts of the system not directly related to the system being changed. In the left-most example, changes to system A only affects system A. In the right-most example, changes to system A affects all other systems because all other systems either directly or indirectly depend on A.

What do I mean by the upfront cost of automation being proportional to the amount of technical debt? An example: Imagine a manual process in which a human being copies files from folder A to folder B in relation to a software deployment. In your manual process, you may without much consideration for the implications of doing so, decide that when you deploy software, you should remember to edit files X and Y, but not Z, because some special conditions apply when you are copying files for deployment to TEST but not LIVE. This is something you may write in a wiki-article, and it will take no more than five minutes to do so. But you cross your fingers afterward that whoever is in charge of deploying your software product next time will read and respect the guidelines outlined in your updated wiki-article. It might be that the activities for deploying to TEST aren't exactly the same as the ones you would execute when deploying to LIVE. Maybe the file structure of your TEST environment isn't the same on TEST, because on your TEST servers there is no D-drive, unlike the servers hosting your LIVE environment. Maybe the operating systems aren't equally patched with security updates, or the database server on TEST is a different major version than the one running your LIVE systems. There are numerous examples, and I imagine you could come up with a few yourself if you have any experience operating software on legacy systems.

Complexity increases cost because you will need to handle and test all exceptions to the default. It isn't enough to validate a deployment to TEST, because the variance when deploying something to TEST uses a different path through your deployment scripts and activities than the deployment of LIVE. In effect this means that the only reasonable way of testing a LIVE deployment is on the LIVE environment. Variance incurs a penalty on non-functional parameters, such as maintainability and reliability, because the complexity increases with the number of exceptions from the default. In other words: Variance increases the likelihood of errors that you may experience.

Variance equals complexity, and complexity equals risk. Variance increases the risk that you as a human being might select or edit the wrong file because it's not the same file that should be edited for all environments. Variance increases the risk that you forget to run a set of scripts in the right order because the order of events aren't the same across environments. Variance increases the risk of failing a build half the time because two build servers aren't configured the same. You may add to this list if you like.

Therefore you and your team should strive to keep all your environments uniform and equally aligned at all times. Do this to avoid handling exception after exception to a simple expectation of where files should be located in your systems, just to keep using that example of making a simple operation unnecessarily complex. "Equally aligned" shouldn't be interpreted in such a way that if you have four similar nodes, i.e. webservers, responding to user requests on your LIVE environment, then you should have four completely similar nodes on all nonproduction environments as well, regardless of the type of traffic they may respond to. You may very well defend having only two webservers in your testing environment, but every single node should be equipped identically to a node on your LIVE environment.

It may be okay to deviate slightly, in terms of something like available disk space, because the amount of traffic that a TEST environment responds to doesn't generate very much data on your hard drives. But decisions like that should always be an exception to the rule of equality and based on a well-founded rationale. If your database instance operates with logfiles on an L-drive, so should all other systems and nodes in your infrastructure. If your webservers on LIVE have a D, E, and F drive, so should everybody else across environments. If you have dedicated hardware for multiple solutions, you should refrain from bundling responsibilities together and having multiple services running on the same hardware on your TEST environment. Laziness or a lack of licenses are in no way acceptable counterarguments for not aligning your systems appropriately if you have the chance to pay down technical debt as described above but shy away from doing so.

In organizations with lots of legacy, the entire infrastructure may have been a series of snowflake implementations over a series of many years, even decades, and version control and templating for various reasons have not been built into processes and organizational culture of "ways of working." In these cases, you should expect to find many deviations from environment to environment. Variance is inevitable in infrastructures generated over a long period of time by many different people with variances themselves in competences and skillsets, which enables them to absorb and adjust their implementations according to decisions made by others. The price of quick and dirty will always have to be paid, and it often comes unexpectedly during times of automation projects. This can send the cost of automating through the roof. In effect, variance in infrastructure and a lack of documentation have the potential to harm a company's ability to compete in its market. It is very hard to achieve speed and agility in an organization dealing with all the exceptions that always need to test against when developing new solutions.

If variance (V) equals complexity (C), and C equals risk (R), then you can write the following formula:

$V = C = R$

...or in short: $V = R$

In other words: Variance equals risk!

Remove exceptions to the default to remove ambiguity in processes as well as in software implementations of business requirements.

Keep it simple—keep everything aligned. This is the single best advice I can give you in terms of improving the quality of a legacy infrastructure where deviations and exceptions are the norm and where data quality in testing environments are mediocre at best. My experience from fifteen years in the business tells me that the root cause of most major outages happens due to lack of quality in the testing environments, rendering it impossible or very time consuming to prepare and execute a trustworthy test of new core functionality with lots of intertwined dependencies.

These conditions will not be solved by automating software release management, since this debt it not only technical. It is also a symptom of debt in your organization and your processes that should be addressed first and foremost by the senior leadership team. An organization should pay down technical debts continuously in the same way that it reorganizes and adjusts business processes. Paying down technical debt by spending time refactoring code or replacing legacy technology is no different from replacing a machine that has been worn out due to extended use throughout the years.

The problem is, however, that machines and buildings exist in the real, physical world, and people can understand that something probably isn't right if a machine is leaking massive amounts of oil or doesn't respond when the green "Start" button is pressed. Timely refactoring exercises in legacy codebases looks exactly the same as creating new customer functionality though. In both cases you just see developers sitting in front of their multiple screens doing stuff you can't relate to. It's impossible for outsiders to determine whether or not a full-featured team on any given day of the week is working on fixing the oil leak, replacing the machine altogether, or building new revenue-generating features.

Automation may help you as a leader to generate knowledge and ammunition to build awareness about the current state of your environments. Automation may help enlighten people and decision makers, but it's not the ultimate solution if there is a gap in understanding organization-wise that variance equals complexity and complexity equals risk. If this is the case, you would rather, as a development team and as a team leader, focus your efforts around communicating how variance and dependencies penalizes you in day-to-day operations. If your organization does not recognize this or simply does not have the skillset required to absorb the level of abstraction required regarding the implications in terms of risk when having to deal with lots of legacy, it is your job and your first priority to build a shared language together with your stakeholders that allows you to have that conversation. You are the ones who know something that others do not. You may hypothesize around the cost of automation up front and then set up scenarios in terms of cost if you approach automation from different angles; regardless, the drivers of this conversation must be those who understand the nature of the problem, taking the business priorities and strategy into account while doing so.

Required Steps in a Deployment Pipeline

Let's look back for a bit at Team Fun Dog's current deployment process as it was outlined in Part 2.

As described earlier, we will use the notation that a green activity is one where the team is in full control over execution. The team has the necessary access to systems and servers for configuring the system, copying files from A to B, etc. Everything needed to successfully execute and adhere to all entry and exit criteria of an activity belongs with the team.

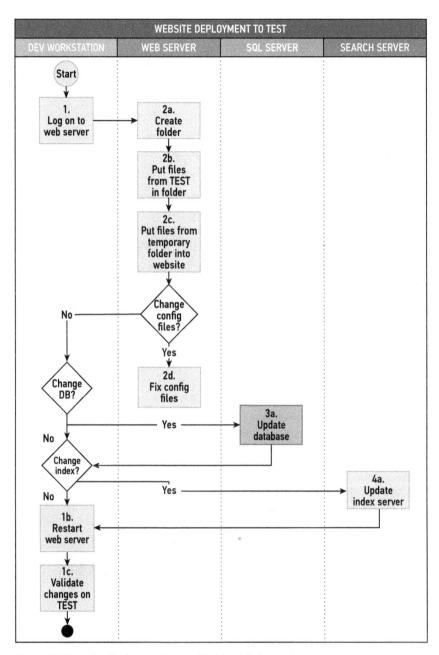

Figure 26. Team Fun Dog's current, manual build and deployment process.

A red activity means the team has a dependency, so it is not in full control over execution. In the example above, Team Fun Dog has a dependency on another department, Henry the database administrator. He alone, as a representative of IT operations, has write access to the production database. For historical reasons, Team Fun Dog does not have this access and therefore must align with Henry in order for a deployment to their LIVE environment to be successful.

In the chapters to come, we will take a deep dive into the activities needed to plan and execute when deploying software. At the end I will spend some time investigating a few solutions for how Team Fun Dog may render this dependency obsolete. There are many ways to overcome dependencies, and I will give you a few examples on how I would approach such a situation. It is my hope that the proposed solutions may inspire you to figure out how to handle similar dependencies in your context at your employer.

We should begin with the beginning. It all starts by downloading an artifact.

Retrieve Artifact

The first step is to retrieve the artifact you wish to deploy, enabling you to cherry-pick the elements and files you need to deploy on the exact node you are working on right now.

Once you download your artifact onto single nodes, it is best practice to always retrieve the artifact directly from your artifact repository and not from another environment, as depicted in Figure 27. This is an antipattern I have seen more than once over the years.

It is by far better that all environments download the artifact from the same location.

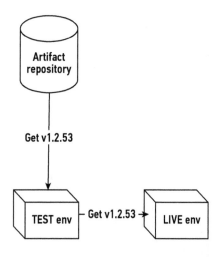

Figure 27. Antipattern when retrieving artifact for deployment on environment. TEST downloads the artifact from an artifact repository, but LIVE does not follow the same pattern.

The difference may seem insignificant, but the devil lies in the details, as we already know. Complexity equals risk. You and your team commit the crime of forcing a set of unnecessary dependencies into your deployment process by engaging in a construct that follows the antipattern outlined in Figure 27. The most important one being that:

You and your team have just elevated your TEST environment to be a prerequisite for a LIVE deployment.

What you are effectively doing by configuring the deployment pipeline in a way so that it retrieves the artifact from TEST is setting up an unnecessary entry criteria for a LIVE deployment. This entry criteria states that your TEST environment should be operational for a LIVE deployment to begin.

Let me ask a rhetorical question now: Would it be fair to assume that the expected operational stability of your TEST environments is not similar to the expectations for operational stability in your LIVE environment? I would say probably. The follow-up questions for what the motivations behind coupling TEST and LIVE environments so tightly could have been would likely breed some quite interesting discussions and realizations individually as well as on a team level.

In most cases where you and your team share infrastructure with other teams, there is always a risk that somebody else on a different project running independently from yours does something to your TEST environment. They could introduce changes that have nothing to do with the deliveries and services you and your team are accountable for delivering. Nonetheless, when that happens, you are unable to test and validate your deliveries due to the instability of TEST forced upon the environment by a completely different team. It's not your fault; you might not even be aware that other projects were deploying changes to TEST at the risk of shutting down the entire environment for everybody.

If you depend on TEST being available because you have built in a dependency that really isn't necessary, you and your team might not be able to deploy to LIVE as planned. Deployment would have been entirely possible had you only retrieved the artifact from the artifact repository instead of relying on a system with no or very loose service level agreements. It has suddenly become much harder to be the best person you can be that day if you have no possible way of delivering the promises you've made to the business for reasons well beyond your span of control. You could have mitigated that risk a long time ago by designing your deployment pipeline without unnecessary dependencies on latently unstable infrastructure.

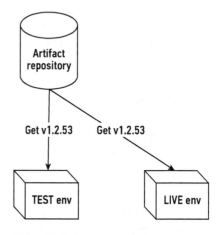

Figure 28. Reduce variance by having all environments retrieve the artifact from the same location always.

There is also the odd case of having IT operational staff and software developers fixing stuff directly on TEST because they know that changes to TEST are mirrored to LIVE by design. One could imagine a team of developers and operational staff experiencing weird bugs but not having the time to follow the correct process of using the Continuous Integration and Delivery platform for building and deploying a fix. They might fix the error on TEST and quickly deploy a fix to LIVE afterward manually.

That works well, unless you forget to put your fix into version control as well. If you forget, you'll just introduce the error the next time you deploy anything to LIVE, and believe me, this will happen if it in any way can, despite best intentions. There is no road shorter to mistrust and pissed-off end users than reintroducing errors repeatedly. If you design your systems and processes in a way that allows for this to happen, even theoretically, rest assured it will someday.

The solution to mitigate all this is to let LIVE deployments download artifacts from the artifact repository always, just like TEST does (see Figure 28). It is a simplification that mitigates the risk of fixes that are not in version control being introduced and leaking from one environment into another. It would be an interesting exercise to mine for conflict as to why someone would choose to fix anything on TEST rather than fixing the problem the right way by building and deploying a new version. In my experience, behavior like this tells a story of gaps in performance or inadequate testing opportunities in the early stages of development, e.g., on your local development machine. There is gold to be uncovered concerning inner and outer motivational factors if you invest an hour or two into a postmortem of outages or incidents where errors were reintroduced because changes only made it to an environment but not into the version control system itself.

In my experience, developers don't do this intentionally. It is more often a matter of trying to work around the penalty of having to wait for a build and deployment to finish. That's a fair counterargument. It is by all means important that end users don't experience downtime any longer than they absolutely need. I would only add that the diagnosis is correct, but the cure is wrong! You do not want a fast lane to getting hotfixes, which is a solution some teams might think is the right way to go. It is not. You should have only one deployment path, one single way of ever deploying software to your systems. If you invent multiple ways of delivering the same artifact to LIVE, you will end up circumventing all automated quality assurance validations by not running all of them. You will not be able to document which tests have passed successfully and which ones you have ignored either by design or by error.

If the build and deployment process takes too long, you should solve that problem, not work around it. You should focus on being able to deploy changes quickly without paying a penalty in terms of longer lead times and increased complexity in your deployment design. You will need to understand, absorb, and maintain several ways of deploying changes to LIVE.

Maybe you remember from earlier in this book where I elaborated vividly about the importance of speed in your build and deployment pipelines? I figure that you have a more profound understanding of why speed matters. Lots of problems simply vanish if your pipelines are able to execute quickly. You may have other challenges then, but it won't be because you haven't run your entire test suite.

Variance equals complexity, and complexity equals risk. Strive to have only one deployment pipeline for any given digital delivery. A discussion about how to split up an artifact into smaller pieces so that each can be delivered more quickly or how to parallelize running your big, automated test suite does much more for end user quality than trying to figure out meaningful ways of short-circuiting essential gates that validate quality.

It's crucial that you force stakeholders to check in changes to your version control if they want to introduce a change to your environments.

Always retrieving the artifact from the same location is a pattern that will help you implement this kind of behavior in your teams. That's why it is important to always follow this procedure regardless of the type of digital delivery you're deploying to your environments.

Prepare Your Environment

In all your work, you should embrace a strategy of "fail fast" in the sense that you should strive for feedback and validate assumptions often to get valuable input and adjust strategy continuously. This applies to the leadership team assessing whether or not investments into high-risk projects should continue or be shut down because a hypothesis has proved itself to be wrong. Fail-fast methodologies also apply to software craftmanship itself, where you should build code to fail fast if an unexpected condition that cannot be gracefully handled arises in the flow of execution. In the case of build and deploy pipelines, you want to probe and validate assumptions prior to beginning something like a high-risk operation of changing state on a node. Validate the state of your environment and fail the deployment altogether if the current state doesn't satisfy expectations.

You may perform a series of low-key validations, such as ensuring that the correct version of a software package is installed on a node. If not, it makes little to no sense to perform an update if you know for sure it will render the node useless afterward, unable to respond to requests from users. You might want to validate third-party dependencies, such as an API that your new version about to be deployed depends on. If the vendor has stated that a new version of their API is up and running at the time of your deployment, which you have scheduled in advance, and it turns out that it is not up and running or that it is experiencing an unexpected outage, you really want to avoid deploying anything that will incur poor quality in the end user experience for your customers.

If you and your team aren't in full control over your infrastructure because you, say, deploy changes to nodes that aren't "yours," then you and your team cannot avoid having dependencies that incur risks you need to mitigate. Team Fun Dog, which happens to have a dependency to Henry the database administrator, is a good example of such a dependency on other parts of their organization.

Let's assume that Team Fun Dog deploys software to servers, which in this case would be nodes with a self-contained operating system such as

Windows Server or a Linux distribution. The server may run on-premise or in the cloud—that's not important—and the location of the server itself is transparent to the users logging in. They just have a name or an IP address connecting them with the node in question. Team Fun Dog doesn't have access to create servers, configure a network, and so on; that's the responsibility of IT operations. A server is something you get by filling out a template and creating a ticket for IT operations. Once it has been resolved, you have a server that you can log in to.

So far, so good. It would be a good idea to have settled on clear agreements between stakeholders by now. What is it really that Team Fun Dog receives from IT operations, and who does what going forward?

When you have ambiguous interpretations about what the word "server" means to somebody in IT operations versus somebody in IT development or marketing, there is a built-in risk of teams and individuals in various areas of the organization becoming misaligned in terms of expectations for deliveries. Sadly enough, this is a pattern I've encountered multiple times, especially in large enterprise companies. There's usually an on-premise infrastructure that crosses organizational boundaries, and they have different mental depictions of what constitutes a server. If you aren't aware of how language abstractions may be interpreted across different organizational silos and perhaps different workplace cultures, you will encounter quite a lot of friction when communicating with other departments.

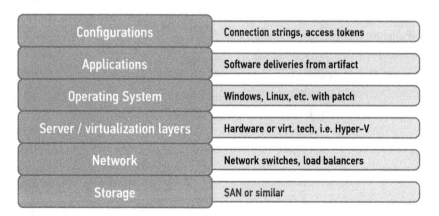

Figure 29. Generic Server Model (GSS): A generic "server" consists of several layers building on top of each other.

Despite having seen throughout the years the fallacies of not paying attention to built-in assumptions when translating seemingly similar words into your own context, I've also found that it's actually quite easy to bridge the gap between different stakeholders applying different meanings to the word "server." Inspired by the OSI/ISO model,[17] I've invented a model of my own that I call "generic server stack," or GSS in short, describing a server as an independent container consisting of several logical layers of abstractions (see Figure 29).

At the bottom of the GSS model, you have storage of some sort. A large hard drive where you can store state and data must be the starting point. On top of this, you can configure network layers consisting of specialized hardware such as a firewall, a DHCP server providing IP addresses to servers, and other means of enabling communication between hardware in an infrastructure.

Now you can add necessary hardware serving different purposes to different people. You may devise network segments for Wi-Fi-enabled devices to connect to the internet; you may have a corporate network connecting the employer's workstations allowing them to print to printers. Different network segments will serve different purposes. You will probably also put virtualization technologies into use that allow you to create servers in your infrastructure on top of raw hardware tailored to host virtual server instances and where you may install an operating system of your choice.

Going even further up the stack, once you have servers available to you with operating systems installed and patched, you can install applications on top of such a server, i.e. a Trend Micro antivirus solution, an Apache webserver, the latest .NET framework, or similar. Let's say you install a webserver. You will then be able to configure this webserver with files from your artifact and configure it to access database instances for data retrieval as the final step in the pursuit of being able to serve traffic requests from end users.

I use the notations in Figure 29 to describe to teams and individuals in areas such as IT operations who are delivering so-called "servers" to their stakeholders. It has served me well as a tool of kickstarting the dialogue between a software development team and IT operations in terms of who

---

17 "OSI Model," Wikipedia, last edit July 13, 2020. https://en.wikipedia.org/wiki/OSI_model

is responsible for what. The likelihood of being able to work out pragmatic solutions increases dramatically when you discuss opinions based on the same depiction of reality. I have come to the conclusion that simple models such as Figure 30 quite often are capable of breaking the ice and building a foundation of shared language that allows for constructive dialogue.

In my experience, most organizations where teams are not in control of all layers in a GSS stack define their responsibilities more or less like Figure 29.

It is the responsibility of IT operations to configure persistence layers, the network layers, everything regarding virtualization, as well as the installation and security patching of an operating system. The IT development team, such as Team Fun Dog, is then responsible for the installment of artifacts and the subsequent configuration of those artifacts. That's the rule of thumb in many organizations that share responsibilities across infrastructure.

In reality the image is much more blurred than the few, graceful, square boxes in Figure 30 seemingly express. Who is accountable for the correct installation of an Apache webserver, for instance? Is it the development team or IT operations? What about when the webserver needs to be upgraded—is that a task to be executed by IT operations or Team Fun Dog? It's a source of frustration in both camps if you don't have this discussion up front regarding who does what given a set of well-defined

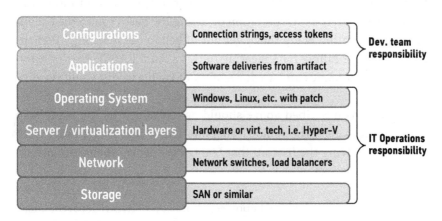

Figure 30. Most organizations where development teams do not have ownership over the entire stack divide the responsibilities of each layer like this.

activities, such as installing and maintaining the necessary toolset and programs that reside on top of an operating system enabling the server to respond to traffic from end users.

More often than not, the IT development team wants to be allowed to do this themselves in their own cadence, but IT operations may have reservations of their own and rightfully so. Just as IT operations may not know anything about Apache webserver setup, how to configure it properly, and how to set up a website serving PHP and JavaScript requests, Team Fun Dog may be just as ignorant regarding code of conduct when patching operating systems, the ins and outs of virtualization technologies, and how to monitor the health of a set of servers taking all systems in the enterprise into account.

It inflicts friction and damages the organization as a whole when responsibilities are divided between different branches of a large organization without agreements between stakeholders aligning expectations about who does what in relation to operational activities once a server has been delivered. In the worst cases, you will find that basic mistrust between departments is the root cause. Each department engages in storytelling about the incompetence of other parts of the organization. When I as a consultant have heard and seen this kind of behavior and then go on a fact-finding mission for myself, I've seen more than once how far away from reality these narratives can be. It hurts organizational performance when nobody makes the effort to apply a bit of interest in the opposing party's interpretation of "doing the best I can to the benefit of my organization every time I get out of bed in the morning." Perhaps something like calling for a meeting to carve out a RASCI chart of activities to assign names and accountability of each activity going forward would help.

Behavior as described above should be addressed by the leadership team, perhaps even the senior leadership team. That aside, I favor two models for solving the friction that teams experience when responsibilities are shared:

1. Agree on a model based upon collaboration between departments.
2. Grant the development team accountability for all layers in the GSS model.

> Structural problems in your organization become visible when you and your team are preparing your environments prior to deployments and you live in a context of shared responsibilities.
>
> The primary indication is that seemingly small and simple tasks capable of being executed in fifteen to thirty minutes end-to-end may spend days or weeks in a queue waiting for somebody to pick them up and execute them in other parts of the organization.

The first solution can be implemented in many ways. Maybe the development team should have a fast track to a person or a team in IT operations who can help them whenever they need assistance. It may work well in some organizations where there is a bit of good will and sufficient levels of trust between departments on a leadership level. You should respect the fact that other teams and colleagues in other departments may have other priorities than yours. The importance of a software project that IT operations staff are working on potentially may benefit the organization much more as a whole compared to the software product that a team such as Team Fun Dog is working on. They may provide value for only 5% of all customers while IT operations is working frantically on implementing GDPR, where the stakes are much higher. Good communication and mutual respect determines whether or not you are able to implement pragmatic solutions, I believe.

Other organizations may have a harder time working more or less informally across organizational boundaries. Not because the directly implicated stakeholders don't want to, but in my experience, enterprise companies happen to inflict bureaucracy to the extent that processes and systems alike grind everything to a halt. Let's take an example such as time registration. It's not impossible at all to find organizations where meaningful collaboration between departments is obstructed by having a resource in department A that is not able to work on a project in department B because employees in department A cannot register time spent on projects that reside in department B. If you obey processes like that and never question them, you will never be able to have people working across organizational boundaries. Bureaucracy within an organization has been implemented at some point for a reason of its own. It should be challenged if there are side effects of legacy processes effectively diluting a stakeholder's desire to do good and be the best corporate citizen they can be every day when they get up and go to work.

You should, in my opinion, by default aim for a model where the development team one way or another becomes capable of assuming full control over the entire server stack. Full-featured teams perform best over time on axes such as fewer defects in produced products and the speed of delivery from business demand to actual delivery to customers.[18]

Another reason that I want to highlight this is that if teams in your organization suddenly become end-to-end responsible for the entire lifecycle of a software delivery—from ideation to delivery and monitoring in production—friction is removed by design. When I see organizations where teams are only responsible for parts of the lifecycle such as development, operations, or a project management office (PMO), people will suboptimize processes. I've seen development teams getting priority when they create tickets for IT operations because they are working on priority projects.

IT operations support might be told that they must prioritize tickets coming in from Team A—the development team—at the expense of Teams B, C, and D. This type of behavior is suboptimization in its purest form, because when resources are scarce, behavior like this just inflicts further delays in other parts of the organization as a whole. You might just cause Team B and D to miss their deadlines because they estimate a turnover on their tickets of seven days. When team A gets priority, Teams B, C, and D will experience longer turnover on a resolution to their tickets, which causes missed deadlines and ripple effects throughout the organization, which causes marketing and sales departments to reschedule their deliveries. All because IT development couldn't deliver as promised.

If you as a team manager believe that moving bottlenecks around will solve problems, you can be certain that while you may solve problems in one area, you are likely to introduce waste somewhere else in the value

---

18 Nicole Forsberg, Jez Humble, and Gene Kim, *Accelerate: The Science of Lean Software and DevOps: Building and Scaling High-Performing Technology Organizations* (IT Revolutions Press, 2018).

stream. Theory of constraints tells us that once you have identified a bottleneck, you should exploit it rather than work around it.[19] I would not favor an approach of prioritizing tickets as my first plan of attack; I would rather see if the organization as a whole could do something to increase throughput, maybe by ensuring that the quality of the descriptions in the tickets received by IT operations was high enough to avoid wasteful clarification mail threads back and forth between ticket creator and IT Operation staff.

Bottlenecks will always exist in complex organizations, but suboptimization will benefit no one. Even though full-featured teams are a hot topic, they're by no means a silver bullet that will solve all your problems. Nothing really is, I'm afraid. Let's, for the sake of the example, say that your organization has identified that teams not being allowed full control over the entire stack is a problem. Very well, the CEO has assigned a task to his leadership team that he wants them to fix this and implement a solution in the leadership team's respective parts of the organization. The team assembles and decides that development teams should be simply granted permission to create TEST and LIVE environments in a cloud solution owned by IT operations. Each team will get a set of administrative credentials to the cloud vendor selected allowing them to create network segments, database instances, and SaaS solutions, and use whatever technology of choice they desire, be it servers or container technologies. All invoices will be sent to a cost center in the company. It is likely that development teams will be grateful, and job satisfaction in these teams will temporarily increase dramatically.

As a leader and as an organization, you have solved one problem, but you are soon to face a whole plethora of others. How do you control cost when development teams can spin up fifty or a hundred server instances without any kind of background knowledge or visibility into the fact that servers don't by any means come free of charge? How do you ensure as a leader that you, in terms of operations, live up to legal and jurisdictional policies in your company? What about GDPR? Where the hell are the customer data now anyway, when teams can freely decide to create servers in datacenters on every continent on the planet? How do we secure and ensure this and

---

19 Eliyahu M. Goldratt and Jeff Cox, *The Goal: A Process of Ongoing Improvement* (North River Press, 2014).

that? You will, as a senior leadership team, face different scenarios of risks and have to apply new mitigation strategies, Your managerial mitigation of risk doesn't just vanish because the execution of a task will be facilitated in other parts of your organization tomorrow. The risk still exists for things like data leaks, but you will need to rework and implement different mitigation strategies if you fall in love with the tales of delivery cadences from full-featured teams and decide to pursue that value proposition in your context.

You should, and rightfully so, set forth a different framework of expectations for your organization—on the team level but also down to each individual—when teams request more freedom and autonomy. Maybe teams should be asked to make continuous budget forecasts of server costs in the cloud. This will perhaps alleviate the behavior and cost-consciousness that you want to achieve from a leadership perspective.

There are solutions to be found by mitigating risks in different ways, but you should be aware that you are engaging in a transformation project that will affect the daily lives of everybody involved, including your own. It is at times of transformation that your personal abilities and inner motivations are tested to their limits. When expectations for execution and accountability of a task change, it will necessarily result in concrete changes in behavior for the new expectations to be fulfilled.

You can and should push responsibility downward as a leader in your organization. It will likely be the best strategic thing to do in the pursuit of scalability by allowing decisions to be made closer to the source of the problem at stake. Context is king, which is why self-service solutions and freedom within a well-defined framework of constraints is the right answer for scaling organizations of a certain size. Full-featured teams capable of planning and executing activities leading up to the goals of producing business value in their context will lead to happier employees, lower turnover, fewer defects in the resulting deliveries, and a better score in the 360-degree yearly review.[20]

It is in the interfaces between IT operations and the rest of the organization delivering software that you best get to learn the culture of an IT organization. When you prepare your environments without being in full

---

20 Forsberg, et al., *Accelerate*.

control over the entire stack, you will by definition interface with teams, departments, and individuals working on other agendas with a different prioritization ladder than yours. You and your team will have to relate to them and actively investigate how to mediate and mitigate bottlenecks as you go along. Show that you're worthy of their trust, and don't anticipate that others behave like fools because they have consciously decided to do so. Open minds, facts rather than feelings, and a bit of humbleness go a long way—even in the largest and most process-focused organizations—to smooth edges and even out misalignments occurring due to rigid process constraints.

Let's get back on track. This chapter's purpose is to prepare your environment for production. That means you should ensure that your infrastructure and single nodes all currently operate and respond as expected. Keeping in mind the chapters about built-in complexities regarding roles and responsibilities, you and your team have a set of entry criteria for the expected state of each node and it's technical dependencies prior to introducing a state change.

There is a plethora of technologies capable of automating installation and configuration of an operating system and applications residing on top of it.

If you and your team haven't got much experience in automation, I would recommend investigating concepts such as infrastructure as code (IaC) to determine which technologies are the current market leaders on the technology stack you're working on today.

The preparation itself may consist of an installation of missing software or a validation that software packages have indeed been previously installed. It is possible to program your way out of installing most programs using a script or technologies that allow for a declarative description of how you want a node to be configured. There are a number of technologies available where you may code a sort of "recipe" for what you want to have installed on each node in your system. Then you let the technology of choice decide how to change state on your node using software packages. This recipe acts as an abstraction over concrete installations, updates, patching, and configuration of everything installed on a node of any kind, allowing you to have full transparency and traceability over changes to your infrastructure. There are many vendors out there, such as Puppet, PowerShell DSC, Chef, Ansible, and others, but the primary use case is basically the same.

There are numerous advantages to this approach. It allows you to, say, install a new server from scratch if one is handed over to you by your IT operations department without inflicting much pain or time consuming manual installation of software before you can put it into use. It also allows you to set up entire environments on the fly if you want to load test a given feature on a clean set of servers. At some point, You will as an organization will need to treat your infrastructure like a commodity, adding leverage of source control systems and Infrastructure as Code patterns and principles, if you believe in the value proposition of Continuous Integration and Delivery. Even though a team has end-to-end accountability, they will never be able to fulfill their potential if they cannot spin up their own testing environments at will.

Installing a server before deploying your artifact will potentially increase the complexity and lower the speed of execution of your deployment pipelines. You also need to align expectations with other stakeholders, such as IT operations, if you want to automate the installation and configuration of webservers on a shared infrastructure. They would want to be informed at least that you have decided to do so, which allows them to raise their concerns if they don't find it suitable to have development teams install software on servers they consider to be theirs. You probably get the point, having the last chapter readily in mind.

If you and your team are not allowed to complete an installation on the nodes you wish to deploy changes to, or do not have the competences to do so, , you should at least validate the consistency and state of your node prior to a deployment. I'll recommend that you programmatically validate that the component has been properly installed as expected after installation. This validation may just be a series of small scripts that open up a file controlling the existence of a specific text string. A script may investigate whether or not a folder has been created as expected. Let that be an initial smoke test validating that an installation has completed successfully. There is an upper limit for how far you can get by validating installations using fragments of scripts because you don't want to end up writing lots of validation code in procedural code. But rest assured that it's perfectly viable to start out this way to gain experience and dip your toes.

The activity of preparing your environment should fail fast and hard if the state doesn't comply with expectations. Do not try to deploy anything on nodes that don't comply with the standards you have outlined or when dependencies will not be able to service requests as expected. As a starting point, you shouldn't want to design graceful failover into your deployment pipelines. Variance equals complexity equals risk, so keep it simple. Abort your deployment pipeline if the state isn't correct. You will have to diagnose and wait for a fix to be applied until you retry at a later stage.

It's the rule in most companies that you to some extent share infrastructure between development teams. It's not evident that the consequences will always be grave, as sound tradeoffs may have been made on the basics of reasonable arguments, but you need to be constantly aware where you have your technical dependencies on other teams. Worst case, teams are sharing nodes where state changes all the time, such as a webserver in a testing environment. Having shared nodes with lots of things going on day in and day out will from time to time cause teams in one part of the organization to make a mistake or introduce instability that ripples through the entire organization. This forces idle time onto other teams because they are by design intertwined and rely on each other's capabilities to keep systems stable at all times.

The thing is that even though teams want to, they may not be in a position where they can test components in combination with each other due to dependencies and restrictions in the infrastructure that prevent them from accessing data on something like a business intelligence system on their individual workstations. In order to test a delivery, they may be forced to introduce untested software or new integrations onto a shared environment, risking the entire environment becoming unstable and unable to respond to traffic from anyone in the company.

Let's assume for a moment a situation where more teams have automated their deployments onto a set of shared resources, such as webservers on TEST hosting a test version of websites for all teams in IT development. If a team needs to change a global setting on a server during their deployment, such as installing a new software, it might be a good idea to build in a mutex lock[21] to mitigate the risk of having two teams, A and B, execute their respective deployment pipelines in parallel at the same time (See Figure 31). If Team A is changing state globally because it is about to install a new version of a software component on the server, it may cause the deployment of Team B to fail unexpectedly with unforeseen consequences. They may cause havoc Team B's deployment if, for instance, a service is restarted as part of a software installation of sorts while they were running a high-risk data migration. Because the service restarted unexpectedly, they may have rendered their entire data storage useless—at least until they retrieve a backup and roll back changes. The frustrations are very real for teams that encounter devastating experiences like this. They can perhaps find a bit of relief in the fact that it's due to a flawed design inadequate for containing impact of changes to each team imposing a change that they struggle in the first place.

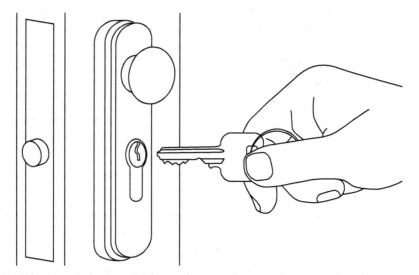

Figure 31. A Mutex lock is basically when you lock something up and make sure that only one key ever exists to get access. A "gatekeeper" role or system then owns the key to unlock your shared resources. Once implemented, other systems can ask the gatekeeper for permission to use the key to get access, one system at a time. Be aware that the easy part is locking up whatever you're trying to protect. It's the processes of acquiring and releasing the single access key that will be the hard part to get right in all situations.

---

21 "Mutual Exclusion," Wikipedia, last edit April 6, 2020. https://en.wikipedia.org/wiki/Mutual_exclusion

One of the strategies you may encounter for mitigation is ensuring that nobody else but you is working on the node you're deploying to. This strategy is not without severe drawbacks, but it may be the right strategy if you apply common sense and do it wisely. If you have no other ways of enabling teams to communicate whenever they are deploying software to shared environments, a shared lock might be worth considering to mitigate the risk of having multiple deployments running at the same time.

It takes a toll in terms of processes to be working with a shared lock between teams. The obvious challenge that comes to mind is to get all teams to respect the mutex lock in order for it to make any sense. That in itself requires that all deployments targeting the shared environment in question have been automated, which means that teams with different priorities and different constraints are all forced to automate deployments before being able to respect a mutual, shared lock. You can already see months passing without getting even close to that level of maturity in your deployment processes.

What if you agree upon manual processes, such as, "The only team allowed to begin a deployment is the team wearing the Deployment Hat"? If you're co-located, it may be the right solution. If it seems banal and a bit childish, it is because rigid processes will never work anyway when performed manually. Humans will interpret those processes differently, and in times of trouble, the human tendency to short circuit means of quality gates will certainly show itself. However, to get you started, employing a banal solution such as wearing an ugly red hat might just be the right solution promoting the desired behavior ("don't deploy unless you have the hat!"). Easy to communicate, easy to remember, and it offers a great means of transparency in some contexts. If you're a distributed team, you might need something other than a physical token serving as a sign of deployment approval, but you get the idea.

Let's assume that everybody has automated their deployments and implemented a means of a shared lock that can be acquired by teams wanting to execute their deployment pipeline. How long are you allowed to block others from deploying to your infrastructure? How do you prioritize who should be granted the lock, and is it reasonable for teams that have acquired a lock for hours to end because they didn't finish in time or experienced unforeseen delays? What if a deployment fails, or you didn't release the lock again? You might decide in favor of a process where an acquired lock is released automatically after a grace period, but what if the team is working and, for reasons outside their level of control, didn't finish, and they need more time—what then? Could you theoretically imagine a situation where a team gets so fed up with waiting that they remove the validation step in their deployment pipeline and simply bypass the process of acquiring a lock before deploying a digital delivery? Believe me, you could very well imagine this situation happening. Within a short period of time, it will become increasingly complex to manage the what-ifs of having a shared lock. As we know, complexity equals risk because something will go haywire at some point.

Despite all its drawbacks, I could agree in certain cases to go with a mutex lock of sorts as a short- to mid-term solution for working around issues with hard dependencies where teams cannot be certain that deliveries and integration endpoints will work as expected. It shouldn't be considered a long-term solution for reasons just described, but it can be useful as pain reliever that buys the organization time to fix the root causes of having shared infrastructure in the first place.

I will *always and without exception* recommend that teams in an organization have dedicated environments of their own, all the way from their development machines to LIVE. The cost of having more servers and infrastructure is an investment well spent in reducing the impact of having one team deploying malicious code and inflicting wait time and reprioritization of work in other parts of the organization. You don't want teams to impose wait time on other teams and impede their ability to deliver to their customers and end users. A strategy of segregating duties along the lines of each team will always be correct from an operational point of view, allowing teams to scale their solutions as they see fit in context of the software deliveries they are producing.

If you still believe that the cost of having unnecessary headroom in terms of seemingly similar infrastructure assigned in parallel for each team you have, consider the cost of having five to eight people fixing an error they have introduced to their TEST environment. Imagine for a second that if they share infrastructure, you suddenly have maybe thirty or forty highly skilled IT technicians and developers constrained from churning on code and deliveries in their contexts. The direct costs are one thing. Never underestimate the indirect costs in terms of follow-up meetings, reprioritization of backlogs, frustrated employers, and deadlines slipping, which forces other departments to also reschedule and miss their deadlines because they didn't get deliveries as expected from IT.

You cannot avoid this entirely, but if situations like this happen with regular frequency, you as a leadership team will find that your best people will quit for a better job elsewhere. They don't feel they are capable of being the best corporate citizen they want to be when they get out of bed every morning when they feel constrained in delivering business value due to the low organizational maturity of their employer. There will always be turnover in an organization—and there should be—but it's very expensive to onboard specialist roles to replace the ones leaving the company. This is why onboarding is also a factor you will need to account for. The number of dependencies and the software architecture as a whole correlate with job satisfaction for software developers and project managers. Don't take my word for it—go ask software teams in your organization how it feels to not be able to provide a trustworthy estimate on even simple tasks or how it feels to introduce three new bugs every time you release a hotfix.

I won't be the judge of right and wrong in this regard; I'm just staging different parts of an overall algorithm and chipping in with my own experiences as a software developer and a DevOps consultant. You as a leader or an employee trying to build a business case for a burning platform must fill in numbers that make sense in your context.

Deploy State Changes

The activity "deploy state changes" covers the change of state on your nodes—such as copying files into a new location once the node has been validated to be eligible for a subsequent deployment. It may be that you replace files with new, updated ones from the artifact you have downloaded. It may be that your deployment pipeline executes a series of scripts from your artifact targeting other nodes, such as a database instance, updating the scheme of the database by adding new columns to a table or inserting updated master data into specific tables. Your state change may consist of a file upload to a file share, meaning that the file is now available for consumers, such as human beings or other software systems operating in another context. The sequence of activities and the content of each may differ greatly, but overall this is where you impose a change to your nodes in your environment.

As we touched on earlier, there may be a long series of sub-activities, each dedicated to a specific task in relation to changing state on individual nodes. Often you will have to shut down services or similar before you change the file system to ensure that files aren't locked by the operating system, leaving you with no options for overwriting it and replacing it with your updated version. A deployment of Team Fun Dog's website on a webserver might consist of three sub-activities:

1. Stop webserver process.
2. Copy files from artifact into website root folder.
3. Start webserver process.

Each of these activities may be self-contained steps in Team Fun Dog's deployment pipeline, but they should all be considered part of an overall umbrella of activities meant to deploy a state change.

Rollback Scenarios

Depending on the amount of traffic and the level of ambition in terms of operational stability, you may have more than one node serving requests, meaning that your deployment pipeline will execute the same sequence of steps on all nodes, including stopping the webserver process, copying files, and starting the webserver again. Imagine a situation where there are four nodes in the form of webservers serving requests; let's name them A, B, C, and D. Webservers A, B, and C complete the sequence of activities successfully, but webserver D fails to do so for unknown reasons. You need to consider how you wish to proceed in the case of only having partially updated your environment.

You have options depending on the nature of each system and the type of change you're deploying. You may choose to roll back to a previous version. This may be cumbersome because it requires you to always prepare and test rollback scripts that are only used in the unlikely event of a rollback scenario. If you deploy changes to database instances that offer a means of transactional consistency when deploying your change, then you can build your change scripts in a way that automatically rolls back changes in the event of a failure.

In my experience, though, it is legacy systems and hardware on which you wish to deploy file state changes where you will most often end up in trouble because file systems on operating systems do not have built-in transactional capabilities. You may change one single file as an atomic operation, but if you change 199 out of 200 files and fail changing the last file, you and your team cannot simply choose to roll back all previous changes to the 199 files in question. There are differences in terms of what individual Continuous Integration and Delivery platforms offer in terms of feature set, but it's important to realize that you will not get any help whatsoever from your operating system when working at the file system level. It's here you are likely to experience errors and incidents due to things like a file being locked by another process. Or maybe you don't have write access to update folders and files as expected, leaving the entire node in an inconsistent state with limited or no abilities to serve requests from users until state has been properly recovered somehow.

My advice on this matter is to design your solutions in a way so file operations alone is a matter of copying a file from A to B in order to minimize the amount of possible errors that may happen. If you use the example of minification of a JavaScript file again, it's not at the time of deployment you want to compress anything. You and your team should instead design your build pipeline in a way so that changes are being carried out and tested as part of your build pipeline, leaving you to only copy a file from the artifact drop folder on your node to its designated target folder on the node being deployed to. The only exception to this rule applies to replacing securities such as passwords or similar, which is a topic we will spend time with in the last part of this book.

Deployment to A, B, and C succeeds, but D fails or unknown reasons. You can choose to fail your entire deployment altogether or simply allow A, B, and C to serve traffic while taking out D for maintenance, meaning that it won't receive any traffic from the load balancer.

It's not hard to program this kind of behavior into your systems, but I would have a few reservations against this approach. For starters, I'd ask how many webservers you can afford to lose. A decision made by someone in the past has revealed that four webservers rather than one, two, or three would be sufficient, so what's the price in terms of operational stability to continue operating your systems with only three servers? There has to be a risk somewhere, so I always recommend that if a deployment doesn't succeed on all parameters, it should fail altogether. In my opinion, it is poisonous to engage in mitigating errors that happen as part of a deployment because it increases complexity and forces you to take lots of what-ifs into account. If you're well aware of the risks and complexities embedded in going down that road, be my guest. It may also be that your Continuous Integration and Delivery platform of choice provides you with some means of built-in capabilities for rolling back changes depending on the nature of what you are trying to achieve. In any case, you should always watch out for unnecessary complexity and fail fast unless everything went as expected.

Changing state and configuring your deployment afterward are by far the most risky activities you will build into your deployment pipeline. All other activities are either preparation tasks or validation tasks, except for the publication of the outcome itself. But errors in these tasks do not inflict any changes regarding the state of the node in question. You should do everything you possibly can to prepare in advance, test as needed that you have the desired state prior to state changes, and then build a simple process of executing the state change itself, keeping complexity and variance to a minimum.

Deploy Configurations

After having successfully changed the version of your digital delivery by copying files and such, you want to configure your digital delivery—if it hasn't already happened implicitly by copying files over and such.

Often you will have some work to do in this regard in terms of configuring your release if you need to access systems such as a database instance using a username or a password—or perhaps an API that requires you to access it using an access token. You may have to extract these credentials authenticating your requests and replace them after you have deployed state changes onto your nodes because each credential is likely to differ from environment to environment. I won't go into depth with secrets here, but I would like to point out that global changes should be applied in one go, and environment-specific changes should be considered a configuration of your deployment in context of the environment you're deploying to. Both activities consist of changing state, but the nature of the changes themselves is different. The former should execute the same operation regardless of environment, and the latter should take the environment itself into account when changing state.

You likely have options on how to handle secrets such as API keys and database passwords. It is a foundational principle to never store your passwords in cleartext inside your artifact. Just don't do it. If you have no policies in your organization regarding how to handle secrets such as usernames and passwords, let this be your guiding star and the design pattern of choice ensuring that you won't leak secrets in the event that somebody achieves unauthorized access to one of your artifacts. I can tell you from experience that artifacts have a tendency to be downloaded to multiple locations when debugging, such as temporary folders on a developer's or IT technician's hard drive. I can tell you from personal experience that if you can live by a simple rule of not storing passwords in cleartext inside your artifacts, you're better off in terms of security than lots of other companies out there.

Run Automated Tests

After a deployment and a successful configuration of your node, you now have the opportunity to run automated tests validating that your node is ready to take traffic.

An automatic smoke test may consist of a few simple checks validating that a service responds as expected. The tests you execute here should to the greatest possible extent simulate how an end user interfaces with the system, meaning that your smoke tests should reside in the very top of the V-model. It makes no sense executing your suite of unit tests or integration tests after deployment testing individual components in combination. You should have validated those tests much earlier anyway.

It's not enough, either, that you validate that a distinct service is running on the machine. It isn't a validation; it only indicates whether or not a service is able to serve requests. You want to actually *use* your node from an end user's perspective to see if everything works as expected after state has changed.

I cannot emphasize enough the importance of executing end-user tests on individual nodes after you have finished deployment of state changes and configurations. If you do not validate each node in a multitier setup individually, you risk unknowingly introducing a periodic error that may be devilishly hard to find afterward. If the same request to a system results in different responses every time depending on which node responds to a user's request, some unfortunate users will experience an error while the rest will get what they requested. Imagine a situation where you are able to log in successfully four out of five times, but the fifth time you get an error because one out of five servers failed the deployment without you noticing because your smoke tests didn't invoke tests on each node. Always, without exception, you should execute all smoke tests at this stage on all nodes if you have more than one.

The tests themselves do not have to be complex, on the contrary. You may have a simple test checking that the webserver responds when you load the front page. If you're clever, you'll embed the version number into your HTML in something like a comment or a hidden field, rendering it invisible for an end user but it nevertheless allows you to explicitly validate that the node is working on the correct version of your software as you expect. It's a simple test—easy to implement, easy to maintain, and it provides invaluable feedback after state has changed.

If there is ever to be a time for scripted user-interface tests in technologies such as Selenium, this would be it. If you and your team raise the level of ambition a bit, you may create a series of automated acceptance tests working at the highest level of the V-model. These tests could validate that your login routines work as expected, that your search engine provides meaningful search results from an end-user's perspective, and that your most precious user paths—converting leads to customers—work as expected after a release. It is a complex exercise to build and maintain such a suite of tests, but the benefits of having verified after each and every release that the ten most wanted and most valuable features of your entire website work after a deployment cannot be underestimated. It's a token of proof when stakeholders seek evidence that their investments into automation and DevOps have paid off.

Just be extremely careful and aware of the maintenance penalties in maintaining a test suite like this. If you want to succeed in having lots of tests, you will be facing a refactoring of the product itself for it to support being automatically tested in the first place. Both tests and the user interface of the delivery under test are likely to be refactored repeatedly over time to support small changes or exceptions to the rule, just as legacy software components often need to be refactored multiple times to support automatic testing.

I would recommend starting out small with the absolute minimum number of tests you can defend. These tests should be no more than smoke tests validating what you think you already know. Ensure that the basics work from an end-user perspective and limit yourself to no more than a handful of tests to begin with.

Publish Result

Regardless of whether or not your deployment went well, you want to advise relevant stakeholders about the result of your state change. In my opinion, there can be only two messages to be shared: it either worked or it didn't. The complexities of "Well, it kind of worked" will blow up in your face quicker than you know it. Your deployment either went perfectly or it did not. In any case, relevant stakeholders need to be made aware somehow.

You may jump over the fence at the lowest point and let it be "good enough" for starters. It's sufficient in the beginning that people can log in and see for themselves if everything went well. If the platform offers features for enabling self-service notifications, such as an email, you may advise people to use this feature for starters.

Deployment Information Should Always Be Relevant

I would recommend though that you do yourself the favor of investing a bit of energy into the publication of a deployment to your stakeholders, especially after a LIVE deployment and/or a LIVE release. Publishing the result of your build is likely to only be of interest to the team committing changes to the source code here and now. In terms of deployment and release, the need to push timely information to stakeholders is of much greater importance, depending on which environment you've deployed an update to. Stakeholders outside the team have no interest in being notified that a new version has been built, but if a new version has just been deployed to LIVE, rest assured that it will be of relevance to other parties besides the team itself to get that information served on time and on a silver platter!

A leadership team should feel that they've gotten their money's worth after approving budgets for an automation initiative. I believe that it is up to the delivery team implementing the automatic deployment pipelines to understand what information is relevant to their stakeholders and how that information should present itself to be perceived relevant by the potential receivers. It may be important to focus on providing the senior leadership team with a dashboard of sorts, but if you ask me, dashboards tend to be expensive to build and provide little value in the long run. I would suggest instead focusing on stakeholders committed to the delivery experience itself without being actually involved. Focus on ways to feed information out into the organization so they are provided with relevant, automatic feedback from the systems when state changes occur.

Let's take Greg from support, for instance. He will be able to answer swiftly in the case of a failing deployment going on when a customer calls up support and asks why the website is responding so slowly or the smartphone app is taking forever to log in. There's a lot of quality in it for Greg to say, "I can see one of our teams is deploying a new version of our login system this very instant. I'll check up with them to ensure that everything is okay." That's far better than saying, "I'm sorry for you having this poor experience. I'll check up on it and return your call as soon as possible." The customer at the other end of the line will in my opinion feel much better knowing that the employees in support know what's going on. Greg could throw in a white lie and say that he can see somebody working on the login system without knowing that to be true, but it makes Greg frustrated when he has to apologize for outages without any sort of warning about maintenance and updates happening. It makes him frustrated and imposes stress and frustrations into his day at work. There's a penalty to be paid by someone somewhere in the organization when people are working blindfolded. In this case, it's up to Greg to decide whether it's him or the customer who should pay the price for not having information at his fingertips when he needs it.

You and your team have great opportunities for offloading and helping your organization by either pushing relevant information out or optimally giving them means of self-service capabilities so they can tailor means of notifications about systems being deployed and released. All involved parties, teams and leaders alike, will be perceived as a huge success if you in your systems understand how to flow timely and relevant information upstream to your stakeholders, even if the code quality and performance is only average at best.

Remember the list of stakeholders we were working on in the first part of this book? We investigated how different stakeholders such as Greg the supporter, Henry the database administrator, and Debbie from marketing would benefit from automation and which concerns they might have. Once you've come this far, it's not that hard to design a good stakeholder experience in terms of flowing relevant information out to them. I would suggest pulling out the list of stakeholders and mapping them into a simple stakeholder map, such as in Figure 32. Make one stakeholder map per environment, because the relevance of information changes dramatically from environment to environment.

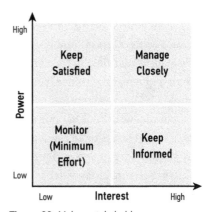

Figure 32. Make a stakeholder map per environment and map all stakeholders according to how relevant it would be to get notifications about state changes for that environment.

You may have development teams that you share infrastructure with who always want to know everything that's going on the TEST environment because they will be able to diagnose outages on TEST much more quickly if they are made aware of other teams working on the shared environment. On the other hand, they might not care at all when your deployment enters LIVE because it's not relevant for them.

Figure 32's "interest" axis is easy to comprehend. The Y-axis, "power," is a bit more subtle. I would suggest that you consider "power" to be people who have power over you. Your primary sponsor, in the form of a senior leader, by default has power over your team. But a stakeholder such as Henry the database administrator's team leader in IT operations, Sarah, should also be considered a powerful stakeholder because her attitude toward automation greatly affects how you are able to implement your automation initiative.

Take all your stakeholders—your leadership team, Greg, Henry, Debbie, product owners, you name it—and map them out in terms of importance. Please don't just decide for yourself if a LIVE deployment is important or not for a given stakeholder. You need to go ask them. They will love you for doing so, without exception.

In parallel with seeking out this information, you and your team should enable some means of self-service feature schemes for everybody. Document how stakeholders can configure mail notifications, Slack notifications, or whichever channel makes sense for people to use. Once you have most stakeholders present in your map, actively investigate how stakeholders with lots of interest and lots of power would want their notifications delivered. If IT operations have a pager duty scheme, they might want a text message that they can automatically forward to the person on pager duty, while Greg the supporter just wants an email delivered in his inbox, which he might be able to configure on his own.

> You and your team haven't tested your communication just because it has been sent to the receiving party. You will be done only when you have confirmed explicitly that information embedded in your outbound communication can be correctly interpreted and understood without assistance.

Information should be timely, relevant, and easy to understand for it to be of any immediate value to a recipient. It's no good to send out the same information to all stakeholders, because IT support might need different information than a member of the senior leadership team. It's your job to ensure timely but also relevant information, meaning that irrelevant information should be left out. This may be my biggest concern in terms of using built-in email templates on Continuous Integration and Delivery platforms because they tend to orient themselves toward a very technically aware audience. It is absolutely *impossible* for a non-tech-savvy recipient to decode what an email subject line such as "Team Fun Dog just released master_v2.3.24 to env LIVE, exitcode 0, timestamp 2020-05-02T14:02:23" really means. It's gibberish if you're not cursed with a lot of technical experience. If you as a CEO want to be notified every time the web shop is updated on LIVE because you want to watch closely how this automation initiative that people won't stop talking about is progressing, and you get a subject line like that, you don't feel any safer or wiser that your budget prioritization has paid off. Regardless of the technical skills employed into your build and deployment pipelines, the perception of quality will be mediocre at best with this communication. At worst, the CEO in this case might think you're incapable of delivering upon the strategy he and his leadership team are trying their best to implement. He is inclined to think like that because you haven't done anything to bridge the gap between technology implementation details and the stakeholder by funding it through a business case with milestones and KPIs. The lack in quality in the delivery of the notification itself will make you look bad. This is a fallacy you must pay close attention to with your most powerful stakeholders.

What's relevant for a CEO is to get an email or a text message stating that "Team Fun Dog released a new version of our web shop to customers today at 2:00 p.m. All automatic tests have passed." With just a little effort and tailored messages, you will increase the likelihood of people actually reading information sent by you and not create Outlook rule-eating notification emails because the content cannot be translated into valuable knowledge and relevance in the eyes of the reader.

In terms of notifying the team in charge of developing the software delivery itself, I would advise in favor of buying licensed software dedicated to improving the feedback cycle back to delivery teams and IT operational staff directly involved with software delivery. There exists a wide range of third-party software for both build and deployment notifications so developers get an integrated feedback experience in the tools they use to develop software. I've worked with solutions where teams install a small piece of dedicated software on their workstations allowing them to quickly notify each other about failing deployments, notify others that they are working on resolving an issue (this is extremely helpful when you are working in distributed teams), and, through the means of dedicated software, take responsibility for fixing an outage or a broken build pipeline. It's essential to complement the team with proper tooling in regards to ensuring fast and easy feedback.

Chat windows are noisy by definition and don't work well for automatic feedback notifications, at least in my experience. I've had much more success with software tailored to solve a specific problems and doing it well, allowing collaborating teams to see real-time status of a deployment without logging in to the platform, being actively notified in case of failing build or deployments, and so forth. I can only recommend that you make a bit of effort investigating what's possible in your context and on your technology stack. The return on investment on a license is nothing compared to being notified and being able to diagnose errors right after they appeared and not hours later.

Summary

If you are working with legacy software and running on aged, legacy infrastructure, you will pay the price during deployment for inadequate documentation and the implications of variance in the landscape of nodes you need to update. There are great opportunities for earning quick wins, especially when you get good at validating your deployment in an automated fashion after a release. Your customers and end users will feel that the quality of the end solution has increased if you become proficient in cancelling deployments or fixing problems before customers even had time to find out.

You shouldn't underestimate the importance of quality communication. Decent code with average performance where monitoring is working well and transparency is high will have a much better chance of sustainable success in the eyes of a sponsor compared to a well-performing, well-architected solution with no monitoring. That leads to angry end users who have to file complaints and escalate tickets before outages are diagnosed and fixed.

Part 5

How to Handle Secrets

IF YOU ARE ACQUAINTED WITH GDPR, you will have encountered terms such as "data classification." In short, data classification is what you do when you define sources of data in your company and assign a risk profile in terms of how sensitive data items and attributes in a data source are in the event of unauthorized access or a data leak. The sensitivity of single attributes on an element in a data source becomes metadata describing the data itself.

Let's take a short example on data classification in practice. Let's assume that a company has a database of customer records containing personal information such as names, addresses, age, email addresses, and perhaps phone numbers. These are very basic pieces of information by themselves, but they are personally identifiable. They can without ambiguity identify an individual human being based on maybe a single attribute, such as phone number, email address, or Social Security number. Companies within the EU aren't allowed to disclose information like this about their customers because EU legislation implemented in each of the membership countries regard this information as particularly sensitive. It should for that reason be handled with extra care by companies and organizations.

Other types of data describing a person, such as sexual preference, religion, health records, and such are also considered sensitive data, but it is in general potentially more harmful or painful for a human to have their entire health record exposed rather than just an email address. It can be devastating to have everybody know about your illness history. It's annoying to have your email address disclosed, but it's hardly heartbreaking and as stressful an experience. Different information describing facets of the same person must be classified differently because the risk profile determining the sensitivity will be different.

Low	Information that are publicly available for anonymous access – such as a blog post that everybody with access to the Internet can download and read.
Medium	Information that isn't publicly available and can potentially be harmful to have exposed. Employment records might belong in this category.
High	Data that could cause serious harm and potentially result in lawsuits against the part abusing the data or who haven't acted with proper care towards protecting access to these data. High-risk data (e.g., journals, mail addresses) may wait.
Restricted	Restricted data are the most sensitive data you can have. A single leak of restricted data may completely destroy credibility with your customers and impede your ability to compete in a market. Credit card information is a good example of restricted data.

Figure 33. Example of risk groups classifying data items in a generic data source.

The starting point in a data classification exercise is to form risk groups such as in Figure 33.

The company should, as part of implementing GDPR compliance, develop its own risk profiles fitting the context of the company. You may have only three profiles that you call "Low," "Medium," and "High" risk. Companies decide for themselves based on a set of GDPR guidelines outlined in the appendixes of GDPR legislation. In terms of GDPR compliance, once a company has implemented risk profiles and assessed their data sources, they must be able to prove afterward, perhaps during an audit, that they store and handle data appropriately for the data that each source represents.

You may save your blog posts in a CMS system for easy publication on your website. This is what you want to do—writing something for everybody to read without requesting anything in return. Employee records should be kept in systems that are protected by a login, and you should be able to tell who has access to your employee records upon request. Different people shouldn't be able to log in using the same username and password. Source code and images of prototypes for nonstrategic product lines may be considered high-risk data.

It's really bad if you leak prototypes to your competitors, but it doesn't necessarily hurt your ability to compete. It may be that you define those images as restricted; that's a discussion you can take up as part of implementing GDPR in your company. If you do so and consider images or details about research and development restricted data, these data items should be encrypted. Even in the unlikely event of a data leak, unauthorized viewers will only get the encrypted data. Without means of decrypting the data, your sensitive data cannot be used against you, your customers, or your trusted business associates who rely on your ability to keep shared secrets safe from prying eyes.

Why is it important to handle secrets? As a software programmer or IT technician, you work with systems every day where you have access to very sensitive data all the time. A leak of data that you work with every day may very well have the potential to threaten the existence of the entire company should that data end up in the hands of people with lots of time, money, and bad intentions.

The pattern, which I am unfortunate enough to see happening over and over again, is that information such as passwords or private keys used to decrypt encrypted data aren't handled with care. Passwords to LIVE nodes in an infrastructure are the most precious and sensitive data you can ever come by when employed within an IT organization. Passwords for other environments, such as TEST, should be considered just as sensitive in my opinion because hackers and friends of hackers are capable of immense patience and persistence if your infrastructure or company brand for one reason or another is of interest to them. The way systems usually get hacked or compromised isn't by leaking passwords out into the open. It's more likely that hackers get unauthorized access to one system and then penetrate system after system, piecing together reams of data from less restricted environments, until they find a weak spot in the defenses of, say, your LIVE environment and start drilling.

This is why a password for your TEST environment nodes poses as much of a threat to you in terms of a data leak as the corresponding LIVE password. Your TEST environment will likely contain data islands and artifacts similar to the ones in LIVE, meaning that it will be a fairly easy for hackers to walk through data on TEST to figure out ways of getting access to LIVE technically or by social engineering. That could call up support and have them reset the password on a user account that they have retrieved details about from a copy of the CRM database restored on the TEST environment.

> If you put passwords in version control without encrypting them, it should, in my opinion, by default be considered a data leak that should have priority and be fixed before anything else.
>
> Why on Earth would you want to save such sensitive data in a versioning control system whose primary use case is one of knowledge sharing?
>
> Get them out. Better now than later!

Most companies are quite good at protecting their LIVE passwords, but the discipline doesn't go much farther than protecting systems used to serve traffic from end users. The plethora of systems used to validate new requirements is often not taken as much into consideration in this regard, even though they often contain fresh copies of data from the LIVE environment.

If you put passwords in version control, it should be considered an incident that has the highest priority to get fixed. The mitigation of a data breach would be to roll the password, meaning that you should replace the password with a new one, rendering the exposed, old password useless to outsiders. That sounds simple enough, but the cost of doing so may vary quite a lot. It can be a simple task that can be completed in ten minutes, or it can evolve into a large project involving the entire organization. It all depends on the nature of your environments, how well architected different systems are, and how they integrate with each other.

If you are using modern, federated solutions to control authentication of your systems so that you don't have to embed passwords into each of the systems requesting data from a given data source, you will likely be able to roll the password without much friction or downtime in the organization to stop the bleeding in case of a data breach. On the other hand, if you have passwords stored in each and every corner of your infrastructure, such as in configuration files in each application requesting data from a given data source, you are in deep you-know-what in case of a data leak. Rather than just changing the password in one single location, such as your active directory solution, you will have to reconfigure every single system in need of data from the data source in question, replacing the old password with the new one.

I've seen examples of companies where an assessment proved that in reality the company didn't have any way of replacing a password in the event of a data breach. Nobody had ever considered this use case. A change of password would require hundreds of nodes to be visited, and each application should be reconfigured—with the built-in risks of changing state. Another fun fact is that you would not be able to change the password and update all systems in one go at the same time. It would be a project of its own to plan a replacement of passwords across all systems. And all this would have to be done without fixing the root cause, so should it happen again, you wouldn't be better off the second time.

It may seem like an impossible mountain to climb, changing a password that has been compromised, if the enterprise architecture and infrastructure aren't properly prepared. Are you working on systems where passwords or similar restricted data will be hard to replace quickly? That's even more of a reason to spend extra time and care on protecting those assets. The price of not paying attention may be disastrous and will position the company badly in the eyes of its loyal customers and in eyes of prospective ones.

The consequences of a breach in GDPR compliance, such as compromising usernames and passwords, can be an expensive encounter. The EU intends, for instance, to fine British Airways a staggering 183 million Euro because "the ICO's investigation has found that a variety of information was compromised by poor security arrangements at the company, including log in, payment card, and travel booking details as well name and address information."[22]

I have no evidence whatsoever, but I'm certain that there have been project managers, system administrators, and team leaders in British Airways who knew that security was scarce and that security shortages were likely to be used with malicious intent if a hacker community got a scent of something worth pursuing. Other companies have been put on

---

22 "Intention to Fine British Airways £183.39m under GDPR for Data Breach," Information Commissioner's Office, July 8, 2019. https://ico.org.uk/about-the-ico/news-and-events/news-and-blogs/2019/07/ico-announces-intention-to-fine-british-airways/

the stand by the EU as well, with the intent from the EU to hand out harsh fines for various types of misconduct in terms of handling sensitive information about its citizens.

You and your team should learn from their mistakes. It is of strategic importance to a company that you do what's necessary in everyday life to ensure a consistent and acceptable level of security. It's important to realize that in reality, it's the software developers and technicians who are the best qualified to assess whether or not security is implemented in a secure fashion. Outsiders may audit and exercise penetration tests to find out how well implemented security is and what to look out for—and companies should do themselves the favor of getting independent sources validating assumptions about their own capabilities in that regard. The fight for privacy starts from the inside, though, and you and your team should as domain experts do whatever you can to ensure the safety of the end user's data.

Simple Is Safer

There's a tradeoff between remaining safe and being able to do work efficiently. You don't want to impose unnecessarily rigid processes or technical constraints upon yourself, such as encrypting data without a valid need to do so or regarding data sources as highly sensitive without the source containing more than medium-risk data.

Fortunately it's not hard to deliver on a security agenda today given the tools and platforms available to us. Ten or twenty years ago, the landscape was very different, with less mature technologies having proprietary or nonexistent integration endpoints out of the box. Most systems today are built with security baked in. GDPR and other legislation have helped a lot in pushing this agenda forward and forcing a change in behavior in terms of consistently ensuring adherence to privacy concerns.

It is, for instance, possible in all Continuous Integration and Delivery platforms that I know of to save sensitive information in ways where information is encrypted without exposing anything in cleartext to outsiders. If it is hard to store and manage sensitive data in the platform you're using right now, you should use this lack in the feature set to consider migrating to another, more mature platform capable of honoring basic use cases in terms of encryption and compliance. You should at this time be able to save a variable and mark it as containing sensitive data that should be encrypted prior to persisting it. It is in my opinion a waste of time and resources to invest in platforms that don't have features like that out of the box.

It may pose great challenges to you if you cannot save sensitive data in your version control systems, because it may prove quite difficult to configure your systems if sensitive data aren't stored as part of your artifact created in your build pipeline. One viable solution is to encrypt the sensitive data prior to storage inside the versioning control system. Encryption allows for a password for LIVE, which in its decrypted form should be considered restricted data, to be considered as perhaps high-risk data instead. Encryption allows for the risk profile to change so that in its encrypted form, you can pass around secrets and other means of sensitive data as part of datasets that are perceived to belong to a risk group of less severity.

Now you have the best of both worlds: a complete artifact containing everything you need, and the necessary flexibility in terms of storing and transporting the artifact around because passwords and such are available in its encrypted form only.

There's no such thing as a free lunch, of course. You will need to decrypt this information somehow at the time of deployment, and that's a complexity you will need to take into account. If you encrypt something, you will need to decrypt it at some point; otherwise, the encryption exercise doesn't make much sense. You will need to build a means of securing the decryption task and securing the private key used to decrypt your encrypted data. If you're using cloud providers, such as AWS or Microsoft Azure, they should all have features and functionality allowing you to delegate the responsibility of encrypting and storing credentials. That leaves you with only the task of providing some neat ways of retrieving a secret when you need it through their API or other secured communication channels available. There are third-party vendors as well, such as HashiCorp, that you may be inspired by and which integrate well into the platforms available as of today.

One thing I've learned throughout my time in the software industry is this: Simple is safer. Make it simple, and then make the simple process safe from unauthorized access. If your pipelines get too complex or your infrastructure is riddled with variance, it becomes hard to secure your systems adequately. The banal, low-key, out-of-the-box solution might be the right one because well-performing solutions, however simple they may seem, are conceptually much easier to comprehend, easier to communicate, and easier to maintain, as long as you don't invent your own encryption algorithms and such. Use built-in features as much as possible in the platforms you have bought licenses for. Security is a solved problem in the tools and platforms available to us today, so don't overcomplicate features and initiatives without explicit need to do so. There are always several ways of achieving a certain level of security, and I'll recommend always going for the easy, well-documented approach rather than the more complex one.

Part 6

Case Studies

How to Handle Dependencies to Henry the Database Admin

Let's revisit the deployment process of Team Fun Dog that we worked with throughout the book and would like to automate as much as possible. The team needs to somehow design a process around Henry the database administrator, who is the person executing change scripts on the database on behalf of Team Fun Dog when they release software on the LIVE environment.

Take a look again at the workflow outlined previously.

Henry is executing scripts on behalf of Team Fun Dog in the red box. Once he has completed his tasks, the team can execute the remaining parts in succession to the tasks he executes.

There are several ways of designing an automated process or at least taking advantage of having a platform that allows for collaboration in a much smoother fashion than when manual labor is coordinated by human beings using emails, chat, or phone calls. In this case study, I'll propose two solutions for working around the constraint of having dependencies to departments and individuals, such as Henry the database administrator

Proposal 1: Manual Approval of Manual Labor

One way of easing the friction of task handover is to ensure that collaborating stakeholders use the same tooling to track progress on tasks. In the case outlined above, one easy way of achieving better transparency and communication would be to design a pipeline with manual approvals of work being done inside the red box.

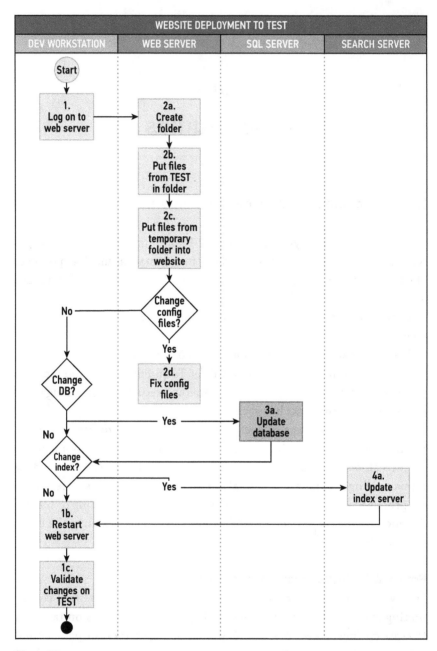

WEBSITE DEPLOYMENT TO TEST

DEV WORKSTATION	WEB SERVER	SQL SERVER	SEARCH SERVER

Start

1. Log on to web server

2a. Create folder

2b. Put files from TEST in folder

2c. Put files from temporary folder into website

Change config files?

No

Yes

Change DB?

2d. Fix config files

No

Yes

3a. Update database

Change index?

No

Yes

4a. Update index server

1b. Restart web server

1c. Validate changes on TEST

Figure 34.

Figure 35. Example of pipeline awaiting human approval. Not until somebody confirms that defined manual activities have completed, will the pipeline proceed.

Team Fun Dog would build in a pipeline where all green activities would be automatically approved in the case of a successful completion of the activity. They would then build in a custom step without activities going on. Then they would configure the pipeline to approve the step manually before proceeding to execute subsequent steps (see Figure 35).

This pipeline will automatically start and finish green activities, but there will be a set of buttons inside the workflow that will need to be pushed by a human being. This kind of manual approval workflow should be readily available in all major Continuous Integration and Delivery platforms of today.

This implies that Team Fun Dog will need to teach Henry how to log in to the Continuous Integration and Delivery platform and either approve or reject his manual execution of scripts once he has executed them. Otherwise, the remaining activities in the pipeline won't be executed.

This change in concrete behavior regarding what is being expected from Henry as part of doing his job is a change management process more than it is about just showing him how it's done. It shouldn't be difficult to explain some of the benefits in terms of eliminating unnecessary communication, having increased transparency of progress state, and so on. Team Fun Dog can up a strong case of "What's In It for Me" from Henry's perspective. He won't need to ask "How long before I can execute the scripts?" He can follow progress just like Team Fun Dog and just do what he's expected to do in a timely fashion without explicit confirmation of whether or not Team Fun Dog is ready. He can approve his work upon successful completion of the steps, allowing the remaining activities to commence immediately.

I would favor this approach unless it would be impossible to automate the process completely, given that there is a valid business case for the investments needed to automate the complexities of getting access to the database and how scripts are executed.

Proposal 2: Automate Manual Labor

If Team Fun Dog is able to automate away Henry the database administrator altogether, this would be the best way if—and that's a big if—there is a sound business case favoring an automation.

If the team always needs to execute database scripts as part of its LIVE deployment, they should automate it if they can. It might seem expensive, but when volume is high, the gains in terms of eliminated tasks and tickets back and forth will soon outnumber the investment. Team Fun Dog should do the numbers and compare the cost of deployment using manual labor for a year versus implementing and supporting an automated solution if they're having difficulty getting buy-in. Team Fun Dog needs to include everything below the visible parts of the iceberg—handoffs, clarification emails, follow-up communication, etc. These all take time that doesn't have to be accounted for when an automated solution has been put into use.

If Team Fun Dog only executes scripts for one out of twenty deployments, meaning that they need to execute a script against the database maybe twice a year, I wouldn't be so sure. A manual, well-documented approach more in line with Proposal 1 may be a better fit for the job in such a situation.

Let's assume that Team Fun Dog and their company all agree that they need to automate Henry's manual execution of scripts, eliminating the dependency all together. Security policies cannot be compromised, however. Team Fun Dog will need to get access to the database from its pipeline without revealing credentials to the database itself.

One way of achieving this is to have the process executing the activities involved in execution of the database scripts run under a service account which has access to the database instance. Team Fun Dog would then build a series of steps in their deployment pipeline that executes scripts placed into a folder in the artifact, against the database in question, see Figure 36.

There is a greater change management job to be done here. Team Fun Dog should not underestimate the importance of spending some time with the stakeholder involved. IT operations may shy away from having systems interfere with LIVE database operations and become defensive. It feels safer to them to have a trusted employee like Henry executing scripts manually rather than having systems do things without explicit approval.

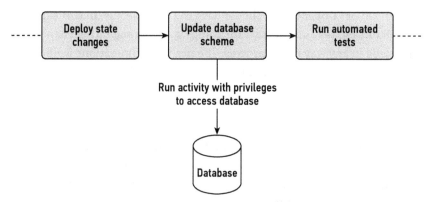

Figure 36. It is possible to design your pipeline so individual activities are executed with different credentials allowing for differented access to nodes of interest.

Team Fun Dog may counter such arguments by trying to say that Henry is likely to be completely unaware anyway about the side effects he is causing on behalf of the team. He isn't the owner nor necessarily a reviewer capable of validating whether or not a script contains subtle errors that could harm the operational stability of the LIVE environment. Chances are that he will execute the script without putting much effort into reviewing the scripts—not because he necessarily doesn't want to pitch in, but because the quality of a review in general relies very much upon knowing what problem you're trying to solve and the reasoning behind the designed implementation details. Review processes should correlate with improved quality in deliveries to end users, so a person like Henry is rarely a good fit for reviewing scripts. He hasn't been part of the process creating it, so the time he spends reviewing won't correlate with improved quality if he hasn't been involved early on in the process. That leaves it open for discussion if his time couldn't be better spent working on optimizing the search experience through improved database indexing for the benefit of everybody, or paying down other means of technical debt on the systems he is in charge of instead of participating in a review theater that has no chance to improve the overall quality of.

When it comes down to it, there are few rational arguments in favor of having humans do repetitive labor, but real arguments doesn't necessarily win people over to your side. Mutual respect does a much better job on that account.

The head of IT operations, should have the right people, in this example Team Fun Dog, take all the heat for things going wrong rather than being implicated in any way in the process. The head of IT operations would help and advise on security regulations and policies in the process, of course. But I would definitely favor an approach where teams in other departments could work independently from IT operations using a platform that offers full transparency into what's going on.

The resources available in IT operations should not spend their time performing tasks such as executing scripts that they are likely to not feel any ownership over. The head of the IT department should instead expect Team Fun Dog to include stakeholders in the process of designing the automated process of delivering changes to the infrastructure that Team Fun Dog is accountable for. The role of IT operations would be one of gating security and GDPR policies, ensuring that things like proper data retention policies are implemented according to company guidelines and that integration points are monitored using the toolset provided by IT operations. These expectations should be set up from the very beginning so that Team Fun Dog knows what's expected from them in terms of non-functional requirements.

It's perfectly fair to stage a conversation this way from a stakeholder's perspective, and it scales very well across teams with varying competences and technology constraints. From an operational perspective, lay out a blueprint and tell people that these are the rules that they need to go by. Then let it be up to them to implement them in ways that take their technology stack and business priorities into account.

There are many variations of proposals 1 and 2; you could combine them in various ways. Just don't underestimate the management process that is buried inside any change in the expectations for how people should do their jobs. You might need to take turns, automating what you can, building in manual approvals where you can't, and then iterating again half a year later when you have earned more trust within the organization and proved to your stakeholders that they can rely on your inner motivations when you argue in favor of further automating deployments in their infrastructure.

It's sensitive stuff to automate configuration of things like a database or a load balancer, so expect stakeholders to have doubts and second guess you. Be prepared for statements such as, "It sounds like a good idea, but IT security policy prevents us from...." It's natural to have these discussions. Push, but don't push too hard at first. Instead of just telling people about benefits and features of automation, show them how a build pipeline works and how risk is mitigated by deploying software automatically. A demo once in a while or weekly twenty-minute check-ins for a month or two goes a long way. You will get everything you want in the end if you and your team understand that you should implement slowly, building on top of previous successes and allowing others to feel ownership along the way.

Tips for Speeding Up Your Build Pipeline

The speed of your build is crucial for getting quick feedback on whether or not recent changes have introduced quality issues that require action. A build should be able to run from start to finish in less than ten minutes—if not, you should optimize the build. Never underestimate the importance of quick feedback. Not getting feedback for twenty, thirty, or forty minutes after changes have been committed shouldn't be considered acceptable performance.

There are many ways to optimize build speed. I'll offer a few tips and guidance for focus areas regardless of technology stack. Before you do anything else, you should time your build so you have a starting point. Devise a table like the one in Figure 37 before you engage into any optimization of your build.

These numbers should be aggregated from the last five or ten successful runs of your build pipeline. It's quite clear in this example that there is something to be found when executing the automated tests, but if you can spend five minutes and reduce "build artifact" time by fifteen seconds, that would be five minutes very well spent too.

Use this simple matrix to map out where to focus your attention and also to get acceptance around the table for what you are trying to achieve. Extend the list of activities as needed and focus around optimizing the sum of all activities. There's no need to set goals for each individual activity. In the example above, it's obvious that the testing phase take the majority of

Step	Time spent (seconds)	Distribution
Download source code	62	~13%
Build artifact	80	~17%
Execute automated tests	321	~66%
Publish artifact	21	~4%
Publish result	1	~0%
Sum	**485**	**100%**

Figure 37. Establish an index 100 by timing your activities first.

time, so you need to do something. Which ten tests take the longest time to run and why? Do you have tests for legacy features that are no longer in use? You can delete these tests altogether. In legacy software, it's more the rule than the exception to have legacy code around supporting obsolete feature sets. You could gain quick wins by cleaning up your business logic and deleting any tests related to this logic that are no longer needed. It will improve speed not only in terms of speedier automated test runs, but in terms of simplifying your codebase. That leaves less room for risk when estimating new features. It's also perhaps a chance to get rid of other legacy code components, because the task of giving old code an overhaul is reduced once you trim the fat and can focus on refactoring code that you know is being used by someone. As a team manager, it will surely also increase team morale when teams understand that it is perfectly okay and even expected that they should spend time cleaning up after themselves and others from time to time.

This waste-reduction exercise will improve the quality of your solution on many essential metrics, not just be beneficial to the lead time of running a build start to finish. And while you're at it, can you run tests in parallel across multiple build agents without inflicting race conditions causing tests to fail? There are lots of viable solutions, but they should all be anchored to a shared goal and explicit, measurable acceptance criteria. That way you and your team have an agreement up front about when you can mark the task of "optimize build" as done.

Manage Dependencies Wisely

An antipattern that I see far too often is when teams put all third-party dependencies into source control without any valid technical reasons for doing so. This introduces longer download time for the source code, because you retrieve both the source code text files and possibly a large bundle of binary files or text files. In reality, version control systems tailored to be used for source code are often poorly fitted to handle large blobs of files. They can do it, of course, but there are subtle penalties you need to pay in day-to-day operations when you include unnecessary files into source control.

Normally when you use a tool such as NuGet, npm, or similar, you only ever need to store a manifest in a flat text file that describes the dependencies you have. This index allows the dependency management tool to build a tree of dependencies and subsequently traverse the tree, retrieve the ones missing, and install them silently without involving the source control system. There's no need to; all you want to know is in the manifest file already. The dependencies themselves can be considered as nothing more than metadata or implementation details for the manifest.

The penalty of downloading and installing packages afterward only needs to be paid once. Caching mechanisms built into the dependency management tools will ensure that downloaded packages can be retrieved afterward from cache folders on your local workstation without downloading packages from the internet every time you build your code. Depending on the nature of the infrastructure hosting your build agents and the technology behind your version control system, this penalty may be more or less severe.

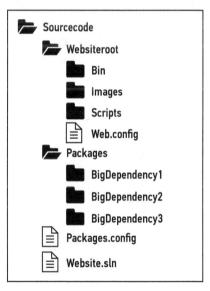

Figure 38. Example from .NET solution—the 'packages' folder with 3 big dependencies should not be under version control, they should be retrieved as part of building the website solution. The packages.config file describing the dependencies should be the only file under source control.

I see it happening over and over again, teams paying unnecessary penalties leading to unnecessary prolonged elapsed build time, which evolved from a lack of understanding about what takes time in their build pipeline. Often, they need to understand and adopt dependency management best practices.

Split Up a Build into Several Independent Builds

In Part 3, where the build process was outlined in detail, I mentioned at some point that you could benefit from splitting up your large artifact into several smaller artifacts. The reason is that having to build and deploy large artifacts in the size of hundreds or thousands of megabytes simply will never become a smooth process for many reasons.

Having one build producing one, big artifact with everything contained in it scales just as badly as it is simple to understand. You may optimize the large build up to a point, but there is an upper limit to how much you can optimize building an artifact if the contents of that artifact will always be of a certain size in order for it to work.

Optimization will at some point need to focus on how you divide your artifact into smaller deliverables and then build in dependencies between them. The example from Part 3 of having an image folder that rarely changes but takes up a lot of space is a good example of where you may apply your split. Creating one pipeline building an artifact that contains all images and another one building the rest of the solution would be one possible solution in this case.

I have personally seen .NET solutions with sixty or seventy projects in them, and I've heard tales from the trenches about solutions with three-digit numbers of projects. Building a single .NET solution with even sixty or seventy projects in it will never perform at any rate, let alone one with more than a hundred. There is a penalty for building a .NET project in a solution. Having that many projects will slow down build time due to the amount of I/O generated by moving around the files compiled.

Having too many projects in a solution is a common antipattern in the .NET world. Splitting up a solution like this into smaller solutions, possibly having a dedicated pipeline building each of them and using dependency management tools such as NuGet to handle that part, will have a positive effect on your ability to build code and deploy code. The challenges of finding your way around a complex hierarchy of projects in day-to-day operations would in itself be less than satisfying for a development team. Being constrained by the structure of the source code greatly affects team velocity and job satisfaction, which is just one out of many valid reasons to consider unnecessary complexities a major architectural defect.

I would look for ways to split monolithic builds into smaller ones if there are no obvious quick wins to be implemented in the existing pipeline and it still doesn't perform well. By having more than one build pipeline creating your combined artifact, it will introduce complexity in your build and deployment pipelines. I believe, though, that when you design multiple

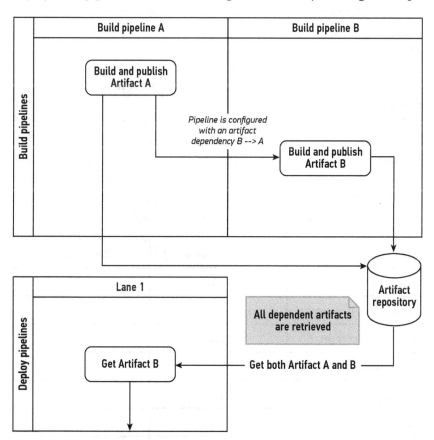

Figure 39. One way of improving build speed is to split up large, slowrunning builds into several, independent ones able to execute quickly and then leverage upon the featureset of your Continuous Integration and Delivery platform to build in dependencies between produced artifacts.

pipelines for building one single whole and apply a bit of common sense when figuring out where to apply the split, the platforms available today are able to handle this type of complexity quite well, which lowers the penalty of complexity in your pipeline.

Tips for Reducing Wait Time in Your Deployment Pipeline

The process of reducing wait time in your deployment pipeline is exactly the same as what you would do when optimizing your build pipeline. As a matter of fact, this strategy is one you would find useful when engaging into any kind of optimization task.

You will need to build your starting point first. In terms of measuring wait time for deployment and for each environment in question, use the table from before and fill in activities.

The numbers in Figure 40 are completely arbitrary. When applied in your context, it would give you an idea about where time is spent and where to get started. You should agree upon an acceptable timeframe of deployment with necessary stakeholders implicated. That's why you have to do this by environment, since the stakeholder map varies greatly from environment to environment, as we have learned previously. Agreeing on a timeframe

Step	Time spent (seconds)	Distribution
Retrieve artifact	28	~4%
Prepare your environment	421	~58%
Deploy state changes	250	~35%
Deploy configurations	2	~0%
Run automated tests	25	~3%
Publish result	2	~0%
Sum	728	100%

Figure 40. Example of measuring lead time on activities in a deployment pipeline.

lets you know both when you are done and when you should react again in the future if elapsed time from start to finish is prolonged beyond what you deem acceptable.

Unlike elapsed build time, which should run for no more than ten minutes as a rule of thumb, it is much harder to carve acceptable wait time ranges in stone. Let's assume you have a deployment pipeline running in a total of eleven minutes. You figure it's unacceptable and want to improve it because as a consequence of deploying new software on this particular system, the software update invalidates an end user's login session forcing them to log out abruptly without prior notice. Due to the nature of the system, end users may lose hours of work when that happens, so the impact of invalidating a user session is severe. That is a source of friction between IT development and other departments in the organization.

You may figure out ways to improve the pipeline so it runs for thirteen minutes but without invalidating user's login session. Would that be a tradeoff worth making? I have no idea; it depends on so many variables. I would without question find that an improved end-user experience would be in favor of extending the acceptable timeframe of a deployment so that thirteen minutes would be okay, but I wouldn't stop looking for ways to improve deployment time. You could get better-performing systems able to swiftly deploy changes without sacrificing logged-in users in the process, but whether or not it is a problem to have deployments running for more than five or ten minutes is a discussion topic regardless. I would favor an approach of having deployments running for no longer than ten minutes. I have learned that sometimes three minutes is easily achievable for a LIVE deployment on one product in an organization, whereas a product in the same organization, but on a different set of nodes with a different set of dependencies, may never get much below eight or ten.

Parallelize Work

If there is one way of quickly reducing wait time, it is by parallelizing work. Parallelization can be quite tricky in a build pipeline, but due to the nature of complex deployments where you need to update many nodes at the same time, you may easily parallelize work as long as each node operates independently in your infrastructure.

Figure 41. Deploying all nodes in sequence could prolong the feedback cycle unnecessary.

Let's assume that you have a setup of a database server and two webservers, in total three nodes in your system. You may execute your deployment on each of them in parallel, like Figure 41.

If each deployment runs on dedicated hardware, such as a webserver A, a webserver B, and a node able to communicate with the database instance, there's absolutely no need to run each deployment in parallel.

It's much better—and should not pose much trouble on modern platforms—to parallelize execution of the three workflows as in Figure 42. When designing your pipelines, you should be sure to utilize parallelization as much as possible in scenarios like this.

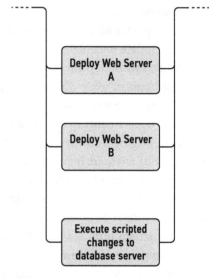

Figure 42. Always consider parallelizing work in deployment pipelines on nodes that operate independently from each other.

You may want to wait to update the database instance until after the webservers have deployed successfully. in case you are unable to roll back changes that you successfully applied to your database scheme after one of your webservers failed its deployment. Always adjust your preferred design so that it parallelizes as much work as it possibly can without compromising the robustness of your solutions in the event that something doesn't go as planned.

Apart from parallelizing work, deciding what to improve will be very context specific. A deployment of containers hosted in Kubernetes has next to nothing in common with a deployment of EC2 instances in AWS built using a combination of scripted and manual installations. Your plan of attack should start by mapping out your activities during deployment and figuring out where you may optimize the most by spending the least effort. Going forward, it's a matter of timing your activities and tracking progress. Rinse and repeat. Keep an eye out for suboptimizations of single steps. If a redesign of your entire deployment would provide much bigger gains for the same effort, it's a discussion worth taking up.

How to Inject Cleartext Secrets into a File During Deployment

It is beyond questioning that you shouldn't store secrets in cleartext in your configuration files. However, legacy systems of various kinds may need to have a password present in cleartext, and this is something you need to work around one way or the other, since you cannot have a configuration file with cleartext passwords in your version control system, such as the example below.

```
. . .
database_failoverenabled: false
database_password: xsF4PF3BAewa
database_defaulttimeout: 10
. . .
```

You don't want this for reasons explained in Part 5. You need to apply a different strategy. A well-established pattern is to replace cleartext secrets from your configuration files in question. Replace your secret with a key that is safe to commit to your version control system.

You would in this situation refactor your configuration file to something like this:

```
. . .
database_failoverenabled: false
database_password: ###DATABASE_PASSWORD###
database_defaulttimeout: 10
. . .
```

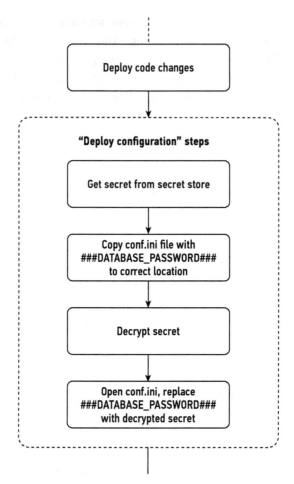

Figure 43. Steps needed to replace secret in conf.ini.

In this snippet, you have replaced your secret with a nonsensitive key called ###DATABASE_PASSWORD###, which you can put into your artifact at no risk of leaking your secrets in the event that your artifact is prone to prying eyes.

Once you deploy your artifact, you will need to build in steps to replace your secrets (see Figure 43).

How you store and retrieve your secret depends heavily upon your Continuous Integration and Delivery platform of choice. Unless you need to have centralized control over secrets such as passwords, I would use encrypted variables so that the platform takes care of both encryption and decryption. If your secret is of a nature where encrypted variables aren't enough or string replacement for one reason or another is insufficient to fulfill the needs you have in your automation efforts, you can design other ways of encrypting and decrypting data by using technologies such as HashiCorp Vault. I would strongly recommend that you only go down that road if you have a very strong business case. The introduction of third-party vendors to handle secrets on behalf of other systems is indeed the solution to secure and align lifecycle management of secrets, but it should be applied on an organizational level and not because one team in a department is building and deploying maybe five or ten software products in total. In this case, the effort and complexities involved outnumber whatever benefits you may have.

Cloud vendors such as Google Cloud, Amazon AWS or Microsoft Azure already offer centralized storage of credentials and secrets on an organizational level. With these vendors, your organization's owners can persist and encrypt secrets while you as a development team or IT operations team can get access by establishing trust between distinct deployment pipelines and the vault containing your secrets. I would favor such an approach if you are unable to meet the needs of your business by simply storing encrypted variables directly in your build and deployment pipelines.

Epilogue

I WILL CONCLUDE THIS BOOK in the hopes that you got what I promised you earlier: Concrete advice on how to automate promoting source code to production and perhaps a bit of inspiration for tackling the challenges you and your team have in day-to-day operations.

As mentioned in the preface, I'm naïve in that I believe people want to do good. In my opinion, it's more often than not mere misalignment, such as flawed communication, that allows friction to induce contradictory behavior between individuals, teams, and departments in organizations of a certain size. Automation and machines may help us in that process if we understand when to instrument machinery to take over tedious, error-prone tasks not well suited for human beings.

Automation is all around us. We have automated tedious chores for centuries, with Gutenberg in the 1400s as a prime example of disrupting established value chains by offering an entirely different way of satisfying customer demand for copies of existing books. We have grown so accustomed to software and automation that we don't see it even when it's right in front of us. Any automobile created today wouldn't be able to start without millions of lines of software being installed in the various hardware components. A washing machine takes a few minutes to load and start, but it would take hours or a whole day to achieve the same outcome doing everything manually. That is the result of decades of finetuning an automated process of washing clothes and linen at different temperatures. Fetch water in a bucket, fetch wood in a nearby forest, make a fire (let's assume for now that we have automated fire-making by having matches available to us), boil water, wash clothes, clean everything up... We see the benefits of automation when we understand how it allows us to focus on activities with a higher fun factor.

Software automation works the same way. When applied well, it has the same potential to ease people's lives, just as a washing machine helps the lives of millions of people every day by automating a process to achieve the same outcome from a set of entirely different activities. It is my hope with this book that you as an ambassador of software automation will do your best to understand which problems automation could solve for your stakeholders in context and apply decent amounts of common sense when implementing automation projects in your software development lifecycle. Following the patterns and guidelines in this book will hopefully help you outline a plan of attack to get you started.

Appendix

Further Reading

Made to Stick: Why Some Ideas Take Hold and Others Come Unstuck,
   Chip and Dan Heath

Starting and Scaling DevOps in the Enterprise, Gary Gruver

Leading Change, John P. Kotter

Accelerate: Building and Scaling High Performing Technology Organizations,
   Nicole Forsgren, Jez Humble, Gene Kim

The Goal: A Process of Ongoing Improvement, Eliyahu M. Goldratt, Jeff Cox

*The Art of Scalability: Scalable Web Architecture, Processes, and Organizations
   for the Modern Enterprise*, Martin L. Abbott, Michael T. Fisher

Glossary

Build
The process retrieving and combining all files, configuration settings and other digital assets needed to update your system at some point. The outcome of a successful build is one or more artifacts.

Dependency management
In software development, you will always build on top of other technologies starting with a freshly installed operating system. This way, you will across technologies always depend on the state and nature of a system for your own code to work. *Dependency management* describes and consolidates the prerequisites for your customized software packages by providing technology-specific means of downloading and installing those dependencies automatically.

Deployment
A change of state on a node in your infrastructure. A deployment doesn't necessarily imply a change in end-user experience.

Digital delivery
A delivery of any digital product or artifact, such as an image, a movie clip, a server installation, a software package, a set of script files, or similar.

Graceful failover
Failover capabilities in a software systems are a means of automatically resolving issues or outages by, for example, starting to serve requests from a different datacenter without impact end user experience and without human intervention. Graceful failover builds on top of failover mechanisms by reducing or shutting down a non-critical feature for a given service to protect core functionality.

Node
A node is a physical or virtual unit on which you wish to perform a state change. A node is likely to be a server, a firewall, a load balancer, or similar, but may also be a database on a database instance.

On-premise

On-premise hardware is an umbrella of hardware owned and hosted by an organization in its own datacenter.

Opportunity cost

If you have more than one choice available to you, each choice represents potential earnings, positive as well as negative. The price of choosing one option over another represents an opportunity cost that you can calculate for each possible option available to you as input to your decision processes.

RASCI chart

A matrix documenting roles and responsibilities on a set of activities you describe that involves more than one stakeholder. An example of a RASCI chart for activities related to a LIVE deployment is in Figure 44. Activities reside on the Y-axis with stakeholders represented on the X-axis.

	Delivery team	1st lvl support	Team manager	Marketing rep.	CIO
Communicate result of deployment to TEST	R		A		
Communicate result of deployment to LIVE	R	I	A		
Release change to customers	S	I	R	I	A

Figure 44. Example of a RASCI matrix useful for documenting stakeholder responsibilities for a given set of activities.

Release

An act in your systems, such as a change in configuration settings, that implies a change in end-user experience. A release may happen independently from a deployment, meaning that multiple deployments don't change end-user experience until a human being performs the release using a process running asynchronously from the deployment process.

Secret

A secret in this book is sensitive or restricted data such as passwords, private keys for decrypting encrypted messages, etc. Secrets are in general to be considered the most sensitive in your organization with the same risk profile as you would place data such as credit card information in, should you need to save these as part of your business operations.

Sunk-cost fallacy

Sunk costs represent money or investments already spent. In economic terms, only future prospective outcomes should be considered when making rational decisions. *Sunk-cost fallacy* describes the irrational human tendency to use large, previous investments as an argument in favor of continuing future investments on top of previous ones regardless of profitability.

Test-driven development (TDD)

A software development process that focuses on writing automated tests as part of delivering features. When you write a feature with a set of automatic tests as part of the delivery, the value proposition is to have Continuous Integration run these tests as part of every code commit, ensuring that delivered functionality works as intended when developers build on top of previously delivered features or refactor existing a feature set to support changing business requirements.

About the Author

KRISTIAN BANK ERBOU has spent more than a decade focusing on software programming, process optimization, and building a culture of collaboration across organizational silos in complex environments. Erbou is an independent Agile and DevOps consultant who helps clients achieve fast feedback loops and delivery excellence in software release management scenarios. He specializes in simplification and optimization of software product lifecycle management (SPLC) and building automation competencies in their organizations. He is a columnist for *Computerworld* and a certified DevOps Institute instructor. Erbou has a clear understanding of managerial motivations and corporate culture based on DevOps initiatives at LEGO, eBay Denmark, and Just-Eat. He lives in Vejle, Denmark, with his wife, three children, and the laziest cat in Northern Europe.

LinkedIn: www.linkedin.com/in/kristianbankerbou
Buildingbettersoftware.com

www.ingramcontent.com/pod-product-compliance
Lightning Source LLC
Chambersburg PA
CBHW052141070326
40690CB00047B/1343